Molecular Modeling and Simulation

Applications and Perspectives

Series editor

Edward Maginn, University of Notre Dame, Notre Dame, IN, USA

This series aims at providing a comprehensive collection of works on developments in molecular modeling and simulation, particularly as applied to the various research fields of engineering. The Series covers a broad range of topics related to modeling matter at the atomistic level. The series provides timely and detailed treatment of advanced methods and their application in a broad range of interrelated fields such as biomedical and biochemical engineering, chemical engineering, chemistry, molecular biology, mechanical engineering and materials science. The Series accepts both edited and authored works, including textbooks, monographs, reference works, and professional books. The series welcomes manuscripts concerned with developments in and applications of molecular modeling and simulation to contemporary research in myriad technological fields, including, but not limited to:

- New Materials Development
- Process Engineering
- Fuel Engineering
- Combustion
- Polymer Engineering
- Biomechanics
- Biomaterials
- Fluid Flow and Modeling
- Nano and Micro Fluidics
- Nano and Micro Technology
- Thin Films
- Phase Equilibria
- Transport Properties
- Computational Biology

More information about this series at http://www.springer.com/series/13829

Randall Q. Snurr · Claire S. Adjiman
David A. Kofke
Editors

Foundations of Molecular Modeling and Simulation

Select Papers from FOMMS 2015

Springer

Editors
Randall Q. Snurr
Department of Chemical and Biological Engineering
Northwestern University
Evanston, IL
USA

David A. Kofke
The State University of New York
Buffalo, NY
USA

Claire S. Adjiman
Department of Chemical Engineering
Imperial College London
London
UK

ISSN 2364-5083 ISSN 2364-5091 (electronic)
Molecular Modeling and Simulation
ISBN 978-981-10-1126-9 ISBN 978-981-10-1128-3 (eBook)
DOI 10.1007/978-981-10-1128-3

Library of Congress Control Number: 2016939101

Printed on acid-free paper

This Springer imprint is published by Springer Nature
The registered company is Springer Science+Business Media Singapore Pte Ltd.

Series Editor's Preface

This is the first volume in the new series *Molecular Modeling and Simulation—Application and Perspectives*. The series aims at providing a comprehensive collection of works on developments in molecular modeling and simulation, particularly as applied to the various research fields of engineering. The goal is to cover a broad range of topics related to modeling matter at the atomistic level and to provide timely and detailed treatment of advanced methods and their application in a broad range of interrelated fields such as biomedical and biochemical engineering, chemical engineering, chemistry, molecular biology, mechanical engineering, and materials science. It is therefore fitting that the first volume contains papers arising from work presented at the 2015 Foundations of Molecular Modeling and Simulation (FOMMS) conference, held July 12–16, 2015 near Mount Hood, Oregon.

I wish to acknowledge the tireless efforts of the FOMMS 2015 conference cochairs Claire S. Adjiman (Imperial College London) and David A. Kofke (University at Buffalo) and conference chair Randall Q. Snurr (Northwestern University), who organized FOMMS 2015 and carried out the editorial duties associated with assembling this volume.

Edward Maginn

Preface

This volume contains ten papers from the 2015 conference on Foundations of Molecular Modeling and Simulation (FOMMS). The theme of this 6th FOMMS conference was Molecular Modeling and the Materials Genome. As in past conferences, the format consisted of invited lectures, contributed posters, and several workshops. A total of 172 people participated in FOMMS 2015, and 116 contributed posters were presented.

The conference began with a keynote address from Frank Stillinger of Princeton University, entitled "Chiral Symmetry Breaking via Computer Simulation." The theme of the first session was Future Trends in Modeling, Simulation and Data Mining, and the session featured talks by Andrea Browning of Boeing, Alán Aspuru-Guzik of Harvard University, and Jinghai Li of the Chinese Academy of Sciences. The session on Biomaterials and Biological Systems consisted of talks from Sabrina Pricl of the University of Trieste and Yiannis Kaznessis of the University of Minnesota. Chris Wolverton of Northwestern University, Kristen Fichthorn of Penn State University, and Jonathan Moore of Dow Chemical spoke in the session on Energy and Environmental Applications, and the session on Complex Fluids and Materials featured talks by Edward Maginn of the University of Notre Dame, Coray Colina of Penn State University, and Marjolein Dijkstra of Utrecht University. Talks by Joachim Sauer of Humboldt University, Daniela Kohen of Carleton College, and Jeffrey Errington of the University at Buffalo were the focus of the session on Catalysis and Interfaces. The session on Reactive Force Fields featured presentations by Susan Sinnott of the University of Florida and Adri van Duin of Penn State University. The conference ended with the awarding of the FOMMS Medal to Carol Hall of North Carolina State University, who gave a memorable talk entitled "Protein Aggregation Simulations: Lessons Learned Over a Decade."

The conference also featured three workshops. The first workshop on Data Mining, Machine Learning, and Materials Informatics was given by Jonathan Moore of Dow Chemical and Johannes Hachmann of the University at Buffalo. Joshua Anderson of the University of Michigan put on a workshop entitled "Using

GPUs for Bigger and Faster Simulations," and the final workshop, entitled "Solving Common Software Problems in Computational Labs," was led by Patrick Fuller of NuMat Technologies and Christopher Wilmer of the University of Pittsburgh.

The principal sponsor of FOMMS 2015 was the CACHE Corporation, with financial support coming from the Computational Molecular Science and Engineering Forum of the American Institute of Chemical Engineers, ExxonMobil, Imperial College London, the Journal of Physical Chemistry, Materials Design, the National Institute of Standards and Technology, the National Science Foundation, Northwestern University, Procter and Gamble, the Royal Society of Chemistry, Scienomics, Springer, the University of Minnesota Nanoporous Materials Genome Center, and UOP.

The ten papers in this volume represent the wide range of molecular modeling tools and applications discussed at the conference. The first paper, by Shao and Hall, presents a coarse-grained model that accounts for protein–protein interactions in a multiprotein system using discontinuous molecular dynamics simulations. The model should set the stage for simulating protein systems on longer timescales and deepening our understanding of processes such as protein crystallization, protein recognition, and protein purification. In the second paper, Sprenger et al. describe their use of molecular dynamics simulations with enhanced sampling methods to study how two types of defects in self-assembled monolayers affect the structure of adsorbed peptides. Moore et al. present the development of a coarse-grained force field for water via multistate iterative Boltzmann inversion. The model is derived to match the bulk and interfacial properties of liquid water. Hülsmann et al. discuss strategies and software for the semi- or fully-automated parameterization of force fields, including options for intramolecular and intermolecular interactions and a work flow combining global and local optimization procedures. In another paper focused on software and methods, Klein et al. describe open-source software called mBuild, which is a general tool designed to simplify the construction of complex, regular, and irregular structures for molecular simulation. Basic molecular components are connected using an equivalence operator which reduces and often removes the need for users to explicitly rotate and translate components as they assemble complex systems. In a methods-oriented contribution bridging quantum and classical mechanics, Subramanian et al. examine the Path Integral Monte Carlo performed with "semi-classical beads." They compare the rate of convergence with respect to the number and type of beads for computing fully quantum virial coefficients of helium-4.

Turning more toward applications, the paper by He et al. describes molecular simulations of the homogeneous nucleation of the ionic liquid [dmim$^+$][Cl$^-$] from its bulk supercooled liquid. Their work combines the string method in collective variables, Markovian milestoning with Voronoi tessellations, and order parameters for molecular crystals. Results include the free-energy barrier, the critical nucleus size, and the nucleation rate. Schweizer et al. study the influence of alloy composition on the structure of Raney nickel catalysts using molecular dynamics simulations and the competitive adsorption of benzene and cyclohexane on Raney nickel as a first step toward modeling the catalytic hydrogenation of benzene. Norman et al. present atomistic modeling related to hydrocarbon mixtures and gas hydrates in porous media, including molecular dynamics simulations to study the phase diagrams of hydrocarbon mixtures in the bulk and in confined geometries. Finally, Bamberger and Kohen report a combination of grand canonical Monte Carlo and MD simulations that provide new insight into an intriguing "cation gating" that allows carbon dioxide but not other adsorbates to permeate Na—Rho zeolites.

We thank all of the participants for their contributions to FOMMS 2015 and especially the authors and reviewers of the papers in this volume. Special thanks goes to the conference facilitator, Robin Craven; the Senior Advisors of FOMMS 2015, Peter Cummings, Joe Golab, Clare McCabe, Jonathan Moore, and J. Ilja Siepmann; and the conference Programming Committee.

Randall Q. Snurr
Claire S. Adjiman
David A. Kofke

Contents

About the Editors

Randall Q. Snurr is the John G. Searle Professor of Chemical and Biological Engineering at Northwestern University. He holds BSE and Ph.D. degrees in chemical engineering from the University of Pennsylvania and the University of California, Berkeley, respectively, and performed postdoctoral research at the University of Leipzig supported by a fellowship from the Alexander von Humboldt Foundation. Other honors include the Institute Award for Excellence in Industrial Gases Technology from the American Institute of Chemical Engineers, the Leibniz professorship at the University of Leipzig, and a CAREER award from the National Science Foundation. He was named a Highly Cited Researcher for the period 2002–2012 by Thomson Reuters. He was a senior editor of the Journal of Physical Chemistry and currently serves on the editorial boards of several journals. His research interests include development of new nanoporous materials for energy and environmental applications, molecular simulation, adsorption separations, diffusion in nanoporous materials, and catalysis.

Claire S. Adjiman is professor of chemical engineering at Imperial College London. She holds an MEng from Imperial College and a Ph.D. from Princeton University, both in chemical engineering. Her research interests lie in the area of integrated process and molecular/materials design, including the development of design methods, property prediction techniques, and optimization algorithms. She is the recipient of several prizes including a RAEng-ICI Fellowship (1998–2003), the Philip Leverhulme Prize for Engineering (2009), and the SCI Armstrong Lecture (2011). She holds an EPSRC Leadership Fellowship (2012–2017) and was elected Fellow of the IChemE in 2013. In 2011, she co-edited a book on Molecular Systems Engineering published by Wiley-VCH.

David A. Kofke received his B.S. in chemical engineering from Carnegie Mellon University and Ph.D. from the University of Pennsylvania, advised by Eduardo Glandt. Since 1989, he has been on the chemical engineering faculty of the University at Buffalo (SUNY), where he served as department chair for 6 years, and now holds the rank of SUNY Distinguished Professor. Author of over 130 refereed publications, Kofke's research currently focuses on rigorous molecular-based free-energy calculations for crystal structure prediction, and calculation of virial coefficients and other cluster integrals from molecular models. Among other awards, Kofke is the recipient of the triennial John M. Prausnitz Award for applied chemical thermodynamics, the Jacob F. Schoellkopf Medal, and the Himmelblau Award from the CAST division of AIChE. Prof. Kofke is a member since 1999 of the Board of Trustees of CACHE, where he served as President in 2010–2012. He is a Fellow of AIChE and AAAS.

A Discontinuous Potential Model for Protein–Protein Interactions

Qing Shao and Carol K. Hall

Abstract Protein–protein interactions play an important role in many biologic and industrial processes. In this work, we develop a two-bead-per-residue model that enables us to account for protein–protein interactions in a multi-protein system using discontinuous molecular dynamics simulations. This model deploys discontinuous potentials to describe the non-bonded interactions and virtual bonds to keep proteins in their native state. The geometric and energetic parameters are derived from the potentials of mean force between sidechain–sidechain, sidechain–backbone, and backbone–backbone pairs. The energetic parameters are scaled with the aim of matching the second virial coefficient of lysozyme reported in experiment. We also investigate the performance of several bond-building strategies.

Keywords Coarse-grained model · Protein–protein interactions · Discontinuous molecular dynamics · Square-well potential · Osmotic second virial coefficient

1 Introduction

Here, we report the development of a two-bead-per-residue protein model that can be used with discontinuous molecular dynamics (DMD) simulations to investigate protein–protein interactions in a multi-protein system. We expect that this new model will allow us to simulate multi-protein systems on longer timescales than what has heretofore been achievable, helping us to deepen our understanding of processes such as protein crystallization [1], protein recognition [2], and protein purification [3].

Protein models can be classified broadly into two types: all-atom if they describe every atom in the protein explicitly and coarse-grained if they group several atoms into one interactive site. All-atom force fields such as CHARMM [4], AMBER [5],

Q. Shao · C.K. Hall (✉)
Department of Chemical and Biomolecular Engineering,
North Carolina State University, Raleigh 27695, USA
e-mail: hall@ncsu.edu

R.Q. Snurr et al. (eds.), *Foundations of Molecular Modeling and Simulation*,
Molecular Modeling and Simulation, DOI 10.1007/978-981-10-1128-3_1

GROMOS [6, 7], and OPLS/AA [8] are very good at describing the behavior of a single protein and how it interacts with other molecules in explicit solvent. However, atomistic simulations are usually limited to one or several small proteins and timescales up to several hundred nanoseconds, effectively precluding the investigation of many interesting multi-protein problems. Coarse-grained models enable us to simulate larger systems for longer timescales using less computational resources. There are two major choices to be made in the development of coarse-grained models: (1) how to coarse-grain the protein geometry and (2) how to obtain the geometric and energetic parameters (see recent review papers [9–15] that summarize the various coarse-graining methods, coarse-grained protein models, and their applications). Coarse-grained protein models can be categorized based on how the atoms are grouped together to form the coarse-grained bead (four-bead-per-residue [16], two-bead-per-residue [17], one-bead-per-residue [18, 19], and ultra-coarse-grained [20]) and how the force field parameters are determined (e.g., Go-type [21], knowledge-based [22, 23], and physics-based [24]).

Coarse-grained models are usually more problem-specific than all-atom models because of transferability issues. Coarse-grained protein models are often developed with the goal of examining particular properties. Most of the current coarse-grained protein models focus on the folding/unfolding problem. It thus remains to be seen how well protein models developed based on such properties do in describing behavior that is a consequence of protein–protein interactions. For example, Stark et al. [25] found that the popular MARTINI force field predicts a second virial coefficient of lysozyme that differs considerably from the experimental value. This inconsistency between simulation and experiment points out the importance of developing protein models that are designed to apply to problems where protein–protein interactions play a major role.

It is also important that the method used to simulate multi-protein systems be fast. Most of the models used in simulating multi-protein systems are based on continuous intermolecular potentials like the Lennard–Jones potential. Simulations based on continuous potentials proceed by solving Newton equations at a uniformly spaced time intervals. They have an algorithm complexity of $O(N\log N)$, where N is the number of particles in the system. The big-O notation describes how the performance or complexity (referring to the number of operations) required to run an algorithm depends on the number of particles in the system. Therefore, the required computational time for continuous MD simulations increases dramatically with the number of beads in the system, limiting their application to relatively small systems.

Discontinuous molecular dynamics (DMD) simulations can be used to investigate large systems efficiently with moderate computational resources. DMD simulations were designed to be applicable to systems that interact via discontinuous potentials (square-well/square-shoulder and hard-sphere). They proceed by analytically calculating the next collision time. Several papers [26–28] describe the details of DMD simulations. The algorithm complexity of DMD simulations is $O(N\log N)$. (One paper by Paul [29] even claims a realization of the DMD method with an algorithm complexity of $O(1)$.) The enhanced algorithm complexity of DMD simulations compared to continuous MD simulations make them suitable for the investigation of long-time processes like spontaneous formation of amyloid structure, which are still challenging for continuous MD simulations.

This work reports our effort to develop a coarse-grained protein model that can be used to study protein–protein interactions in multi-protein systems via DMD simulations. We deploy a two-bead-per-residue protein model: one bead for the backbone and the other for the sidechain. The parameters of our protein model are obtained by coarse-graining atomistic simulation results for backbone–backbone, backbone–sidechain, and sidechain–sidechain interactions in explicit water. The rest of the paper is organized as follows. Section 2 describes the protein model in detail; Sect. 3 describes the atomistic and DMD simulations; Sect. 4 discusses the analysis leading to the final choice of model parameters; and Sect. 5 summarizes the current status of the model.

2 Model Description

We deploy a two-bead-per-residue protein model to represent the 20 natural amino acid residues. Since computational efficiency was a major consideration here, we tried to minimize the number of beads in the system and at the same time represent the chemical heterogeneity of the individual amino acid residues. Although a one-bead-per-residue model minimizes the number of beads in the system, we found that it made it harder to represent the difference among the various types of amino acid residue in DMD simulations. Protein models with more than two beads per residue do a good job of representing the chemical heterogeneity of the 20 residues (see, e.g., our protein model, PRIME20 [16]), but this increases the required computational resources. The two-bead-per-residue model is a good compromise for large proteins.

The 18 amino acid residues except glycine and proline are represented by two beads: one bead at the position of the C-α atom to represent the backbone entity and the other at the sidechain center of mass to represent the sidechain entity. Glycine and proline residues are represented solely by a single bead at the positions of their C-α atoms because either they do not have a sidechain or the sidechain is closely linked with the backbone. Figure 1 shows a schematic of the two-bead model for alanine.

The potential energy of the system is the sum of the intermolecular potential energy, intramolecular potential energy, and virtual bond energy for all the beads in the system (Eq. 1).

$$U_{\text{total}} = \sum U_{\text{inter}}(r) + \sum U_{\text{intra}}(r) + \sum U_{\text{bond}}(r) \tag{1}$$

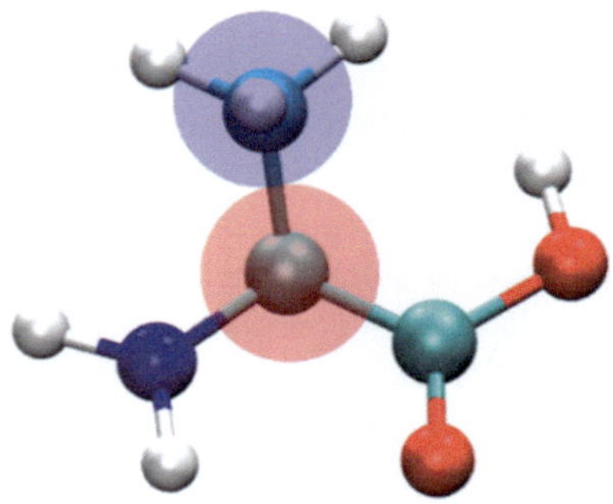

Fig. 1 Schematic of the two-bead-per-residue model. One bead is at Cα, and the other is at the center of mass of the sidechain

The intermolecular bead–bead interactions are represented by a single square well or single square shoulder potential as given in Eq. (2):

$$\begin{cases} U_{\text{inter}}(r) = \infty, & r \le \sigma_1 \\ U_{\text{inter}}(r) = \epsilon, & \sigma_1 < r < \sigma_2 \\ U_{\text{inter}}(r) = 0, & r \ge \sigma_2 \end{cases} \quad (2)$$

where r is the bead–bead distance, σ_1 and σ_2 are geometric parameters, and ε is the energetic parameter. The geometric and energetic parameters (σ_1, σ_2, and ε) are derived from the potentials of mean force (PMFs) of sidechain–sidechain, sidechain–backbone, and backbone–backbone pairs from atomistic simulations in explicit water solvent as discussed in Sect. 4. A single square-well potential ($\varepsilon < 0$) indicates that the two entities attract each other in explicit water; a single square-shoulder potential ($\varepsilon > 0$) indicates that these two entities repel each other in explicit water; and a hard-sphere potential ($\varepsilon = 0$) indicates that the two entities just have an excluded volume interaction in water. The effect of water is taken into account in the parameters because the PMFs were obtained from the pair's interactions in explicit water solvent.

The intramolecular bead–bead non-bonded interactions consider excluded volume effects only. The hard-sphere potential is used to describe the intramolecular bead–bead non-bonded interactions (Eq. 3).

$$\begin{cases} U_{\text{intra}}(r) = \infty, & r \le 0.8\sigma_1 \\ U_{\text{intra}}(r) = 0, & r > 0.8\sigma_1 \end{cases} \quad (3)$$

where r is the bead–bead distance and σ_1 is the geometric parameter in Eq. (2). The geometric parameters could, in principle, be obtained from the volumes of the individual beads, but to simplify the process, we choose to use $0.8\sigma_1$ as the geometric parameter. We found that this selection avoids overlap between beads in a protein and works well with the virtual bond setting, which is described in the next paragraph.

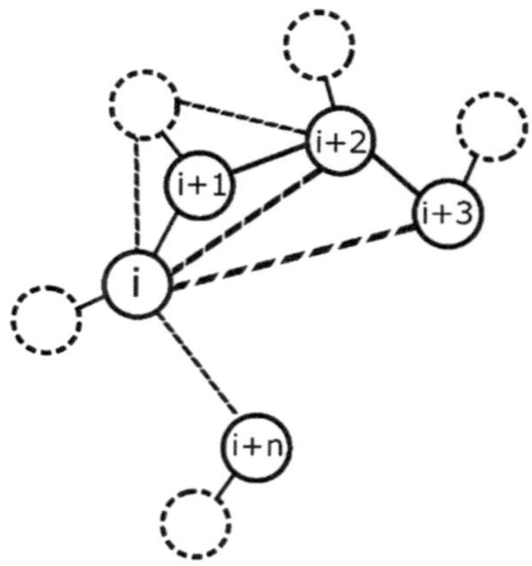

Fig. 2 Schematic describing virtual bonds. The *circles with solid borders* are backbone beads, and the *circles with dash-line borders* are sidechain beads. The virtual bonds connect these beads to keep the protein in its native state

We deploy virtual bead–bead bonds to maintain the protein in its native state. The virtual bond potential is a double hard wall (Eq. 4).

$$\begin{cases} U_{\text{bond}}(r) = \infty, & r \leq (1-\delta)\sigma \\ U_{\text{bond}}(r) = 0, & (1-\delta)\sigma < r < (1+\delta)\sigma \\ U_{\text{bond}}(r) = \infty, & r \geq (1+\delta)\sigma \end{cases} \tag{4}$$

where r is the bead–bead distance, σ is an equilibrium bead–bead distance obtained from the native state of the protein, and δ is the flexibility factor. Figure 2 shows a schematic describing the virtual bonds. The native state of a protein is its naturally folded structure. Here, we use the structure of a protein in the Protein Data Bank (PDB) as its native state. The virtual bonds can be divided into two categories depending on the indices of the connected beads along the amino acid sequence. The "local" category includes virtual bonds between beads whose index difference is less than four. They are used to maintain the protein local secondary structure. The other category (non-local) includes virtual bonds between beads far away from each other along the amino acid sequence. These bonds are used to maintain the tertiary and quaternary structures of a protein. Section 4.2 discusses the choice of the virtual bonds in detail.

3 Simulation Details

3.1 Atomistic Simulation

We conducted atomistic simulations to obtain the geometric and energetic parameters for the coarse-grained beads in the two-bead-per-residue model; these parameters are then used in the DMD simulations. The sidechain and backbone entities were generated from amino acid residues. Glycine and proline entities were generated by capping their N and C terminals with an acetyl group and an N-methyl

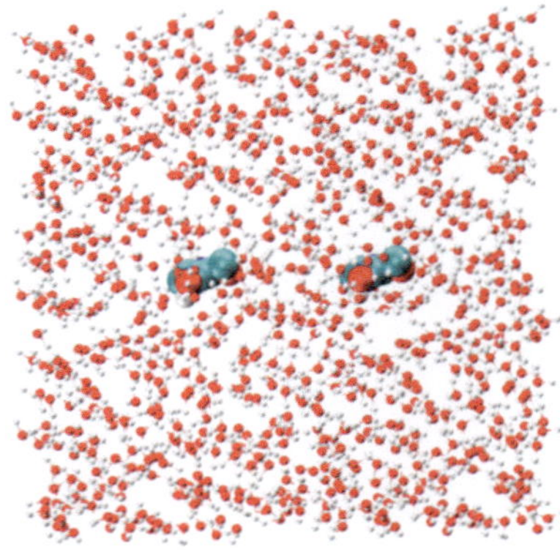

Fig. 3 Glycine–glycine pair in a 3.0 × 3.0 × 3.0 nm^3 box. Glycine molecules are represented in a VDW view, and water molecules are represented in CPK model view l. C atom *green*, N atom *blue*, O atom *red*, and H atom *white*

amide group. These caps prevent the two entities from associating with others through their N or C termini. The glycine entity was also used as the backbone entity because it is an amino acid without a sidechain. Sidechain entities were generated by detaching the sidechain of an amino acid residue from its backbone and replacing the CB atom with an H atom. Two sidechain or backbone entities were placed in a 3.0 × 3.0 × 3.0 nm^3 box filled with water molecules at a density of 1.0 g/nm^3. The initial entity–entity distance was at least 1.0 nm to avoid any artificial association. The GROMOS54a7 force field [7] was used to describe the sidechain and backbone entity, and the SPC model [30] was deployed to describe the water molecules since it is recommended for use with the GROMOS force field. Figure 3 shows the initial configuration of a glycine–glycine pair in a water box.

For each system, a 1-ns isothermal–isobaric ensemble (NPT, $T = 300$ K, $P = 1$ bar) MD simulation with a 1-fs time step was conducted after energy minimization to let the system reach the equilibrated density and potential energy. Then, a 100-ns canonical ensemble (NVT, $T = 300$ K) MD simulation with a 2-fs time step was conducted to collect data every 500 fs. The 12-6 Lennard–Jones interactions were treated with a 1.0-nm cutoff, and the electrostatic interactions were treated with particle mesh Ewald sum [31]. No bonds were constrained to their equilibrium length during the 1-ns NPT MD simulation. The bonds to the hydrogen atoms were constrained to their equilibrium length using LINCS algorithm [32] during the 100-ns NVT MD simulation. The desired temperature was maintained using the v-rescale algorithm [33], and the desired pressure was maintained using the Parrinello–Rahman algorithm [34]. The MD simulations and energy minimization were conducted using GROMACS-4.6.5 [35].

3.2 *DMD Simulation*

We conducted DMD simulations to test and scale the parameters obtained from atomistic simulations. The DMD simulations were conducted in the NVT ensemble

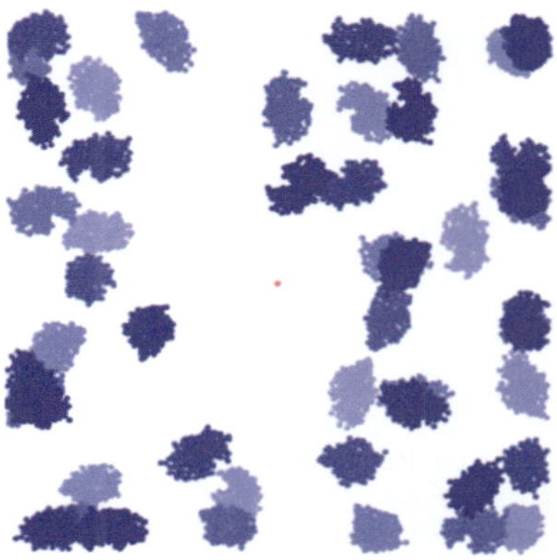

Fig. 4 The initial configurations of 50 lysozymes in a $40 \times 40 \times 40$ nm^3 box (1.38 μM)

using code developed in our group. For single-protein systems, the protein was placed in the center of a $10 \times 10 \times 10$ nm^3 box. The temperature of the system was maintained at 1.0 using the Andersen thermostat [36]. For multi-protein systems, 50 lysozyme proteins were placed at random positions in a $40 \times 40 \times 40$ nm^3 box (1.38 μM) using Packmol [37]. The initial protein–protein distance was at least 7 nm to avoid any artificial association. Figure 4 shows the initial configuration for 50 lysozyme proteins.

4 Parameter Development

4.1 Intermolecular Interaction

We use a pair of glycine (G) entities to illustrate how we get geometric and energetic parameters (σ_1, σ_2, and ε) from atomistic simulation results (Fig. 5). The radial distribution functions between the centers of mass of two glycine entities (Fig. 5a) were obtained from the MD simulation. Boltzmann inversion [38] was used to calculate the PMF (Fig. 5b). There are several ways to select the geometric and energetic parameter from a continuous potential [39, 40]. Here, we choose the geometric parameter σ_1 to be one root where the PMF = 0 (Fig. 5b) and the energetic parameter ε to be the lowest value of the PMF. Here, we choose σ_2 to be where $g(r)$ reaches the range of 1.0 ± 0.1. This method may result in a small energy perturbation (−0.1 kBT when $g(r) = 1.1$ and 0.1 kBT when $g(r) = 0.9$) but avoids the possibility of having an artificially large well/shoulder width when the PMF approaches zero slowly. The geometric parameter σ_1 for the square-shoulder is where the PMF starts to increase rapidly, and the energetic parameter ε is the

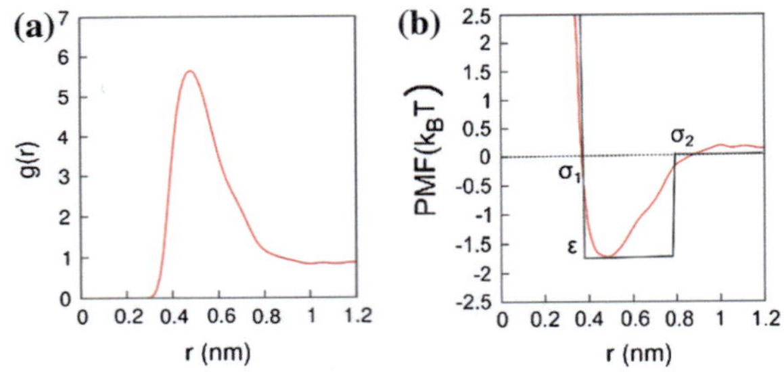

Fig. 5 **a** Radial distribution functions and **b** potential of mean force (PMF) and square-well potential for a G–G pair. The geometric (σ_1 and σ_2) and energetic (ε) parameters for the square-well potential were obtained from discretizing the PMF of entity pairs in atomistic simulations

value of PMF at σ_1. The geometric parameter σ_2 for the square-shoulder is determined in the same manner as for the square-well. If the value of the PMF is always between −0.1 and 0.1 k_BT, we deploy a hard-sphere potential for the bead–bead interaction. The geometric parameter σ_1 is selected in the same way as that for the square-shoulder.

It is of interest to ask whether or not the parameters (σ_1, σ_2, and ε) of the 210 pairs are physically meaningful and if they bias toward certain conformations. The value of the parameter σ_1 reflects how close the two entities can approach each other in water solvent. The majority of bead–bead pairs have σ_1 in the range of 0.33–0.45 nm, which is quite close to the size of the heavy atoms in the entities. These entities should be able to contact with each other directly in water solvent. Only five pairs have σ_1 larger than 0.45 nm: arginine (R)–arginine (R), arginine (R)–lysine (K), glutamic acid (E)–glutamic acid (E), tryptophan (W)–tryptophan (W), and tryptophan (W)–tyrosine (Y).

The value of parameter σ_2 reflects how far apart the two entity beads can be and still influence each other. The values of σ_2 for the 210 pairs range from 0.55 to 1.0 nm. This wide range illustrates the chemical dissimilarities among the 18 sidechain entities. The values of σ_2 for the hydrophilic and charged pairs (such as 0.85 nm for the asparagine–asparagine pair and 1.0 nm for the lysine–lysine pair) are generally larger than those for the hydrophobic pairs (0.55 nm for the valine–valine pair). This is expected because the former two are controlled by electrostatic interactions, which decrease much more slowly as a function of distance than the van der Waals interactions which control the hydrophobic associations.

The value of the energetic parameter ε reflects whether the two entities attract or repel each other. We first consider charged sidechain entities. The pairs of sidechain entities with the same sign charge have positive ε (a repulsive force), and the pairs with opposite sign charge have negative ε (an attractive force). Our atomistic MD

Table 1 The entity pairs that have a hard-sphere potential

G-S	T-R	S-H	R-I
V-D	T-H	D-Q	E-I
V-R	S-R	D-L	Y-Y
V-E	S-K	D-I	Y-W
T-D	S-E	D-M	W-W

simulations successfully capture how these charged entities interact with each other. Histidine (H) has a pKa similar to 7.0, so its net charge is quite weak. Therefore, we do not find a strong repulsive force between H and the negatively charged sidechains. Instead, we find a weak attraction, probably due the effect of water molecules.

We further examine the values of parameter ε for the other entity pairs. Two pairs, glycine–aspartic acid and glycine–proline, have a positive ε, which may be due to their different influences on the structure of water molecules. The other pairs have a negative ε, whose value depends on the chemistries of the entities and their individual effects on the structure of surrounding water molecules. For instance, the value of ε for the valine–valine pair is $-1.44\ k_BT$, and that for the serine–serine pair is only $-0.61\ k_BT$, consistent with the fact that hydrophobic substances associate more stably than hydrophilic substances in water. The glutamic acid–cysteine and lysine–tryptophan pairs have much lower ε than the others. The former may be due to an interaction between the S atoms and the charged group, and the latter may be due to a charged-group-π conjugation.

Twenty entity pairs (Table 1) have a hard-sphere potential because their interactions are judged to be very weak based on the criterion stated above. Some of these may be due to the different hydrophilicities of the entities (such as the sidechains of valine and aspartic acid). Some of these may be due to the effect of water molecules. Consider for instance, the glycine–serine sidechain pair. The serine sidechain has a hydroxyl group, which should be able to associate with the oxygen atom on glycine; however, these two entities can also form hydrogen bonds with water molecules. The water molecules around the two entities may make the glycine–serine sidechain association energetically comparable to the non-associated state. This weak interaction reminds us of the importance of taking the effect of water molecules into account when considering protein–protein interactions.

4.2 *Virtual Bond*

An ideal set of virtual bonds should be able to maintain the protein in its native state, while maximizing the timescale per simulation step. We investigated how this goal could be achieved by tuning the types of virtual bonds and the flexibility factor δ in Eq. (4). Table 2 lists the choice of virtual bond types and the values of δ for

Table 2 Three virtual bond sets. CA[*i*] means the *i*th backbone bead, and CB[*i*] means the *i*th sidechain bead

Local		Non-local	
Bond types	δ	Bond types	δ
Rigid			
CA[*i*]–CA[*i* + 1] CA[i]–CA[*i* + 2] CA[*i*]–CA[*i* + 3] CB[*i*]–CA[*i*] CB[*i*]–CA[*i* − 1] CB[*i*]–CA[*i* + 1]	0.05	CA[*i*]–CA[*i* + 10] CA[*i*]–CA[*i* + 20] (*i* = *i* + 2)[a] CB[*i*]–CB[*i* + 10] disulfide bonds	0.05
Moderate			
CA[*i*]–CA[*i* + 1] CA[*i*]–CA[*i* + 2] CA[*i*]–CA[*i* + 3] CB[*i*]–CA[*i*] CB[*i*]–CA[*i* − 1] CB[*i*]–CA[*i* + 1]	0.12	CA[*i*]–CA[*i* + 10] (*i* = *i* + 2) CA[*i*]–CA[*i* + 20] (*i* = 2, *i* = *i* + 4)[b] CA[*i*]–CA[*i* + 40] (*i* = *i* + 8) CA[*i*]–CA[*i* + 80] (*i* = *i* + 16) CB[*i*]–CB[*i* + 10] (*i* = *i* + 4) CB[*i*]–CB[*i* + 20] (*i* = 2, *i* = *i* + 4) disulfide bonds	0.05
Loose			
CA[*i*]–CA[*i* + 1], CA[*i*]–CA[*i* + 2] CA[*i*]–CA[*i* + 3] CB[*i*]–CA[*i*] CB[*i*]–CA[*i* − 1] CB[*i*]–CA[*i* + 1]	0.25	CA[*i*]–CA[*i* + 10] (*i* = *i* + 2), CA[*i*]–CA[*i* + 20] (*i* = 2, *i* = *i* + 4), CA[*i*]–CA[*i* + 40] (*i* = *i* + 8) CA[*i*]–CA[*i* + 80] (*i* = *i* + 16) CB[*i*]–CB[*i* + 10] (*i* = *i* + 4) CB[*i*]–CB[*i* + 20] (*i* = 2, *i* = *i* + 4) disulfide bonds CA[*i*]–CA[*i* + 120](*i* = *i* + 16)(myoglobin)	0.12 (lysozyme) 0.1 (myoglobin)

The value of $\delta\sigma$ is limited to be less than 0.1 nm

[a]*i* = *i* + *n* means that this type of virtual bonds is set for beads *i*, *i* + n, *i* + 2*n*...

[b]*i* = *n* means this type of virtual bonds starts from *n*th bead

three sets used for lysozyme and for myoglobin. These are labeled as "rigid," "moderate," and "loose" based on δ. The number of virtual bonds in the "loose" category is greater than that in the "moderate" category which is itself greater than that in the "rigid" category.

We select lysozyme (PDB ID: 193L) as our first test protein to evaluate the ability of these three virtual bond sets to maintain the protein in its native state. Lysozyme was chosen as our first test case because it is relatively small and rigid. The virtual bond set's ability to maintain the protein in its native state was measured using the root mean square deviation (RMSD) of all beads from the lysozyme native state conformation during a DMD simulation of 200 million collisions. As shown in Fig. 6, the small RMSD fluctations (0.15–0.30 nm for the rigid set, 0.18–0.30 nm for the moderate set, and 0.15–0.25 nm for the loose set) indicate that all three sets work well at maintaining lysozyme in its native structure. Interestingly,

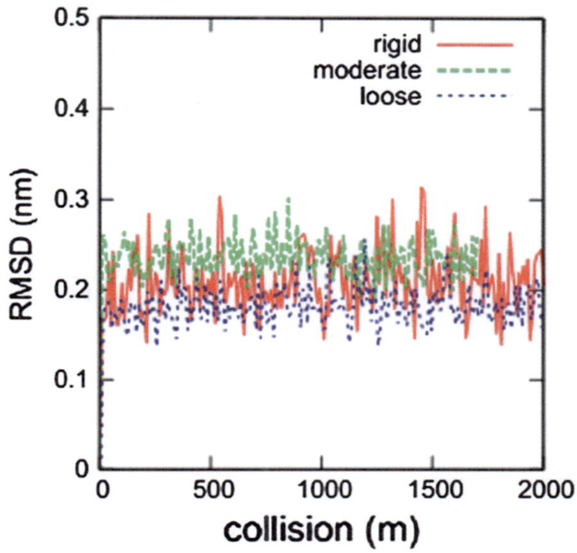

Fig. 6 Root mean square deviation (RMSD) of all beads in a lysozyme during a 2-billion-collision DMD simulation

Table 3 Simulation time advanced by a 100-million-collision DMD simulation of 50 lysozymes

	Reduced time
Rigid	619
Moderate	1457
Loose	2072

the “loose” set works as well as the other two even though its δ is much higher than the other two.

The “loose” virtual bond set works best at maximizing the simulation timescale per million collisions. As listed in Table 3, a simulation of 50 lysozymes shows that for a 100-million-collision simulation, the simulation-reduced time achieved with the “loose” virtual bond set is 1.4 times that with the “moderate” virtual bond set and 3.3 times that with the “rigid” virtual bond set. DMD simulations of complex molecules such as proteins spend more than 90 % of the simulation time in collisions between the bonded beads, indicating that the timescale of a DMD simulation heavily depends on the bond flexibility. A “loose” virtual bond set allows the simulation to advance much faster than a “rigid” one. However, the “loose” set may also increase the risk that the protein will deform. This risk should be taken into account when selecting proper virtual bonds.

We then tested the performance of these three virtual bond settings for a more flexible protein: myoglobin (PDB ID: 1YMB). The RMSD of myoglobin well illustrates the importance of choosing the virtual bond types and flexibility factor more carefully. The “rigid” and “moderate” virtual bond sets for myoglobin are the

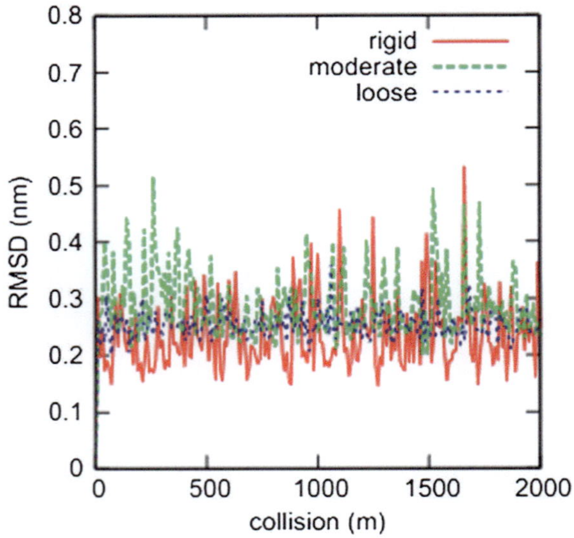

Fig. 7 RMSD of all beads in myoglobin during a 2-billion-collision DMD simulation

same as for lysozyme. As shown in Fig. 7, the RMSDs with these two virtual bond sets fluctuate from 0.15 to 0.5 nm (rigid) and 0.2 to 0.5 nm (moderate). The large RMSD fluctuations indicate that we need to select virtual bonds carefully. We thus set a "loose" set for myoglobin, which has a new type of virtual bond: CA[i]–CA [i + 120] ($i = i + 16$) because myoglobin is larger than lysozyme. In this set, the value of δ for the local virtual bonds is chosen to be 0.25; however, unlike our choice for lysozyme, we set the non-local δ to be 0.10 instead of 0.12. As shown in Fig. 7, the increase in the number of "non-local" virtual bonds improves the ability of the model to hold myoglobin in its native state (0.25 < RMSD < 0.3 nm).

The comparison among the three virtual bond sets shows the importance of the non-local virtual bonds in maintaining the protein in its native state. A protein usually keeps its tertiary and quaternary structures with the help of non-bonded intramolecular interactions such as hydrogen bonds and hydrophobic associations, which are not considered in the current version of our model. All the virtual bonds that connect the beads whose index difference is less than four are there to maintain the bonds, angles, and dihedral angles between the beads. They work well at maintaining the local secondary structure of a protein but have little influence on the tertiary and quaternary structures of the protein. Thus, the model depends on the non-local virtual bonds between beads that are far away in the sequence to hold different regions of a protein together. The selection of "non-local" virtual bonds is still an art. Using the virtual bonds that connect the beads whose index difference is 10, 20, 40, 80 …, up to the total number of the beads in the protein, works well. The value of δ for the "non-local" virtual bonds should be less than that for the "local" ones due to their large equilibrated bond length. Although the total number of "non-local" virtual bonds is much less than that for "local" virtual bonds, the non-local bonds are very important as they greatly enhance the ability of a protein to stay in its native state. For instance, having six virtual bonds between CA[*i*] and CA[*i* + 120] results in a decrease of the RMSD of myoglobin from around 0.2–0.5 nm (rigid and moderate) to 0.25–0.3 nm (loose) (Fig. 7). Such a strategy could be useful for the simulation of other complex systems.

4.3 Energetic Parameter Adjustment

The values of the force field parameters need to be adjusted to ensure that the coarse-grained model gives reasonable results in comparison with experiment. Tables 4, 5, and 6 show the values of σ_1, σ_2 and ε obtained from the PMFs for the 210 bead-bead pairs. Here, we select the osmotic second virial coefficient (B_{22}) of lysozyme as the reference property because it well represents the strength of lysozyme–lysozyme interactions in a solution. Lenhoff and his colleagues [41–44] measured B_{22} for lysozyme in a variety of solutions. Lysozyme is expected to have a positive B_{22} in water because it is positively charged. Their data indicate that B_{22} of lysozyme in water at pH 7 is around 5×10^{-4} mol ml/g^2.

We use Eq. (5) to calculate B_{22} from radial distribution functions g(r) of the center of mass of lysozymes obtained from DMD simulations of our system of 50 lysozymes.

$$B_{22} = -\frac{2\pi}{M_{\mathrm{w}}^2 N_{\mathrm{A}}} \int_0^{\infty} (g(r) - 1) r^2 \mathrm{d}r \tag{5}$$

where M_{w} is the molecular weight of the protein, N_{A} is the Avogadro constant, and $g(r)$ is the radial distribution function. We then compare the simulation value of B_{22} to the experimental one; a simulation value of B_{22} larger than the experimental one would imply that the force field overestimates the attraction among proteins, while a simulation value of B_{22} smaller than the experimental one would imply that the force field overestimates the repulsion among proteins.

The energetic parameters are adjusted so that the value of B_{22} obtained from simulations approaches the experimental value. Ideally, all the geometric and energetic parameters should be adjusted individually, but this would require massive data which are not achievable now. Alternatively, if we fix the interaction ranges of the 210 pairs, i.e., the geometric parameters σ_1 and σ_2, and keep the ratio of the energetic parameters for any two pairs unchanged, we can adjust all the energetic parameters by multiplying them by a single factor *f*. This helps us to narrow the difference between the simulation and experiment results for B_{22}.

Table 4 Geometric parameter σ_1 of 210 interactive site pairs (nm)

	G	A	V	P	T	S	N	D	R	K	E	Q	L	I	F	Y	W	M	C	H
G	0.33																			
A	0.38	0.35																		
V	0.38	0.38	0.40																	
P	0.38	0.40	0.40	0.45																
T	0.38	0.38	0.40	0.40	0.40															
S	0.38	0.38	0.40	0.40	0.40	0.40														
N	0.38	0.38	0.38	0.40	0.38	0.38	0.38													
D	0.38	0.38	0.40	0.40	0.40	0.35	0.38	0.45												
R	0.38	0.38	0.38	0.38	0.38	0.38	0.38	0.40	0.50											
K	0.38	0.38	0.40	0.40	0.38	0.38	0.38	0.38	0.50	0.40										
E	0.38	0.38	0.45	0.40	0.40	0.38	0.38	0.45	0.38	0.40	0.50									
Q	0.38	0.38	0.38	0.40	0.38	0.38	0.38	0.40	0.38	0.38	0.38	0.38								
L	0.38	0.38	0.40	0.40	0.40	0.40	0.38	0.40	0.38	0.40	0.40	0.38	0.40							
I	0.38	0.38	0.40	0.38	0.38	0.38	0.38	0.40	0.38	0.40	0.40	0.38	0.40	0.38						
F	0.38	0.38	0.40	0.40	0.40	0.40	0.38	0.42	0.38	0.38	0.40	0.38	0.40	0.38	0.40					
Y	0.38	0.38	0.40	0.40	0.40	0.40	0.38	0.45	0.38	0.38	0.45	0.38	0.40	0.40	0.40					
W	0.38	0.38	0.40	0.40	0.45	0.40	0.38	0.45	0.38	0.38	0.45	0.38	0.40	0.40	0.42	0.60	0.60			
M	0.38	0.38	0.40	0.40	0.40	0.40	0.38	0.40	0.38	0.38	0.40	0.38	0.40	0.38	0.40	0.40	0.40	0.40		
C	0.38	0.38	0.38	0.40	0.38	0.38	0.38	0.30	0.38	0.38	0.35	0.38	0.40	0.40	0.35	0.40	0.38	0.38	0.38	
H	0.38	0.38	0.38	0.40	0.38	0.38	0.38	0.38	0.38	0.38	0.38	0.38	0.38	0.38	0.38	0.38	0.38	0.38	0.38	0.38

Table 5 Geometric parameter σ_2 of 210 interactive site pairs (nm), "hs" means hard-sphere

	G	A	V	P	T	S	N	D	R	K	E	Q	L	I	F	Y	W	M	C	H
G	0.75																			
A	0.60	0.55																		
V	0.70	0.65	0.65																	
P	0.75	0.65	0.70	0.70																
T	0.65	0.60	0.65	0.65	0.65															
S	hs	0.60	0.65	0.65	0.60	0.60														
N	0.65	0.62	0.70	0.65	0.70	0.70	0.65													
D	0.60	0.65	hs	0.60	hs	0.40	0.55	0.85												
R	0.65	0.65	hs	0.80	hs	hs	0.50	0.80	0.65											
K	0.75	0.70	0.65	0.70	0.60	hs	0.70	0.80	0.80	1.00										
E	0.65	0.65	hs	0.65	0.80	hs	0.60	0.80	0.70	0.80	0.80									
Q	0.75	0.65	0.70	0.70	0.70	0.70	0.70	hs	0.65	0.70	0.60	0.75								
L	0.75	0.65	0.70	0.70	0.65	0.65	0.70	hs	0.70	0.70	0.80	0.75	0.70							
I	0.70	0.65	0.70	0.70	0.70	0.65	0.70	hs	hs	0.75	hs	0.70	0.70	0.70						
F	0.75	0.65	0.70	0.70	0.65	0.60	0.70	0.68	0.55	0.65	0.75	0.75	0.70	0.70	0.70					
Y	0.70	0.60	0.60	0.70	0.65	0.60	0.70	0.80	0.70	0.65	0.85	0.70	0.70	0.70	0.70	hs				
W	0.80	0.70	0.70	0.70	0.70	0.60	0.70	0.80	0.65	0.70	0.80	0.80	0.75	0.75	0.75	hs	hs			
M	0.75	0.65	0.65	0.70	0.65	0.65	0.70	hs	0.75	0.65	0.75	0.75	0.70	0.70	0.65	0.65	0.70	0.65		
C	0.60	0.60	0.65	0.65	0.70	0.65	0.65	0.45	0.65	0.65	0.50	0.65	0.70	0.70	0.70	0.75	0.75	0.70	0.65	
H	0.75	0.65	0.70	0.70	hs	hs	0.70	0.55	0.60	0.70	0.58	0.75	0.70	0.70	0.70	0.70	0.70	0.70	0.75	0.75

Table 6 Energetic parameter ε of 210 interactive site pairs (k_BT), "hs" means hard-sphere

	G	A	V	P	T	S	N	D	R	K	E	Q	L	I	F	Y	W	M	C	H
G	−1.61																			
A	−0.81	−1.44																		
V	−1.21	−1.01	−1.21																	
P	−1.42	−0.81	−0.81	−1.42																
T	−0.40	−0.81	−0.81	−0.81	−0.40															
S	hs	−1.13	−0.81	−0.61	−0.40	−0.61														
N	−0.81	−1.01	−1.13	−1.01	−0.61	−0.81	−0.81													
D	0.40	−0.20	hs	0.40	hs	−1.01	−1.01	0.81												
R	−0.40	−0.20	hs	−0.81	hs	hs	−0.20	−0.81	0.81											
K	−0.40	−0.40	−0.61	−0.81	−0.61	hs	−0.61	−0.81	0.81	0.40										
E	−0.40	−0.40	hs	−0.40	0.40	hs	−0.81	0.61	−0.61	−0.81	0.61									
Q	−1.21	−1.10	−0.81	−1.01	−0.81	−0.40	−0.81	hs	−0.61	−0.61	−0.40	−0.61								
L	−1.41	−1.21	−1.01	−1.01	−0.81	−1.01	−1.01	hs	−0.40	−0.81	−0.40	−1.21	−1.21							
I	−1.21	−0.81	−1.01	−1.21	−0.81	−0.81	−0.81	hs	hs	−0.40	hs	−0.81	−1.01	−0.81						
F	−1.62	−1.01	−0.81	−1.42	−0.81	−0.40	−1.21	−0.41	−1.62	−0.40	−0.40	−1.21	−1.21	−1.21	−1.21					
Y	−1.21	−0.81	−0.40	−0.61	−0.20	−0.40	−0.40	0.61	−2.22	−1.42	0.40	−0.81	−0.81	−0.61	−0.81	hs				
W	−1.21	−0.81	−0.81	−1.21	−0.40	−0.40	−0.61	0.61	−2.42	−2.83	0.40	−0.81	−1.01	−0.81	−0.81	hs	hs			
M	−1.41	−1.01	−1.01	−1.21	−0.40	−0.61	−1.21	hs	−1.01	−1.01	−0.40	−1.21	−1.13	−1.13	−1.01	−0.81	−1.01	−1.01		
C	−0.81	−0.81	−0.81	−0.81	−0.81	−0.61	−1.01	−3.03	−1.13	−0.81	−2.43	−0.73	−1.01	−0.81	−1.01	−0.48	−0.81	−0.81	−0.73	
H	−0.81	−0.81	−1.01	−1.01	hs	hs	−0.81	−1.54	−0.40	−0.40	−1.21	−1.01	−1.21	−1.21	−1.13	−1.01	−1.01	−0.81	−0.60	−0.61

The reduced temperature T^* is set to 1.0 when tuning the factor. For each f, the average value of B_{22} was obtained from three independent DMD simulations starting from different initial configurations. These simulations lasted for 120–170 billion steps with a total reduced time τ of around 1×10^6.

We find that f needs to be small in order to get our value of B_{22} to be close to experimental value. The value of B_{22} is $1.9 \pm 0.83 \times 10^{-4}$ mol ml/g^2 when $f = 0.15$ and increases to $7.1 \pm 3.18 \times 10^{-4}$ mol ml/g^2 when $f = 0.10$. These values are close to the value obtained experimentally (5×10^{-4} mol ml/g^2). We chose to set $f = 0.10$ as the scale factor because the simulation results for B_{22} with $f = 0.1$ straddle the experimental value. It is not surprising to find such a small f value. There are two possible reasons for the need for such drastic rescaling. First, coarse-graining smooths the free energy surface and makes it easier for proteins to aggregate. Second, the current coarse-graining method may not be able to address the effect of water molecules near the proteins well. Stark et al. [25] also found that they needed to drastically rescale the energetic parameters of the MARTINI force field [45] to match the experimental value for B_{22} of lysozyme. The necessity of scaling parameters to match experiment results was also observed for ionic liquids [46]. A possible reason for such necessity is that the current models use an additive two-body interaction system, which is an approximation to the many-body interactions. Parameter scaling may be an effective way to attenuate the error brought by the different interaction systems.

5 Conclusion

We have developed a discontinuous potential two-bead-per-residue protein model so that we can conduct DMD simulations to investigate protein–protein interactions in a multi-protein system. The current model focuses on proteins that are in their native states. We derive the intermolecular bead–bead interactions from the potential of mean force obtained from atomistic simulations. Examination of the geometric and energetic parameters shows that these parameters are physically meaningful. We also developed strategies to set the types and flexibility of the virtual bonds to constrain the proteins in their native state while maximizing the simulation timescale. Comparison of a variety of virtual bond sets illustrates that high bond flexibility (the loose set) improves the DMD simulation performance. We also scale the energetic parameters of our model to match experimental results on the osmotic second virial coefficient of lysozyme. We are using this model to investigate the formation of the corona of proteins that forms around a nanoparticle.

Acknowledgments This work was supported by National Science Foundation (CBET-1236053) and the National Institutes of Health (EB006006). This work used the Extreme Science and Engineering Discovery Environment (XSEDE), which is supported by National Science Foundation grant number ACI-1053575.

References

1. Kastelic, M., Kalyuzhnyi, Y.V., Hribar-Lee, B., Dill, K.A., Vlachy, V.: Protein aggregation in salt solutions. Proc. Natl. Acad. Sci. U.S.A. **112**, 6766–6770 (2015)
2. Azzarito, V., Long, K., Murphy, N.S., Wilson, A.J.: Inhibition of α-helix-mediated protein-protein interactions using designed molecules. Nat. Chem. **5**, 161–173 (2013)
3. Hober, S., Nord, K., Linhult, M.: Protein A chromatography for antibody purification. J. Chrom. B **848**, 40–47 (2007)
4. Best, R.B., Zhu, X., Shim, J., Lopes, P.E.M., Mittal, J., Feig, M., MacKerell, A.D.: Optimization of the additive CHARMM All-atom protein force field Targeting improved sampling of the backbone ϕ, ψ and side-chain $\chi 1$ and $\chi 2$ dihedral angles. J. Chem. Theory Comput. **8**, 3257–3273 (2012)
5. Hornak, V., Abel, R., Okur, A., Strockbine, B., Roitberg, A., Simmerling, C.: Comparison of multiple Amber force fields and development of improved protein backbone parameters. Proteins: Struct. Funct. Bioinf. **65**, 712–725 (2006)
6. Huang, W., Lin, Z., van Gunsteren, W.F.: Validation of the GROMOS 54A7 force field with respect to β-peptide folding. J. Chem. Theory Comput. **7**, 1237–1243 (2011)
7. Schmid, N., Eichenberger, A., Choutko, A., Riniker, S., Winger, M., Mark, A., van Gunsteren, W.: Definition and testing of the GROMOS force-field versions 54A7 and 54B7. Eur. Biophys. J. **40**, 843–856 (2011)
8. Jorgensen, W.L., Maxwell, D.S., Tirado-Rives, J.: Development and testing of the OPLS All-atom force field on conformational energetics and properties of organic liquids. J. Am. Chem. Soc. **118**, 11225–11236 (1996)
9. Tozzini, V.: Coarse-grained models for proteins. Curr. Opin. Struct. Biol. **15**, 144–150 (2005)
10. Wu, C., Shea, J.-E.: Coarse-grained models for protein aggregation. Curr. Opin. Struct. Biol. **21**, 209–220 (2011)
11. Saunders, M.G., Voth, G.A.: Coarse-graining of multiprotein assemblies. Curr. Opin. Struct. Biol. **22**, 144–150 (2012)
12. Baaden, M., Marrink, S.J.: Coarse-grain modelling of protein–protein interactions. Curr. Opin. Struct. Biol. **23**, 878–886 (2013)
13. Noid, W.G.: Perspective: coarse-grained models for biomolecular systems. J. Chem. Phys. **139**, 090901(1–25) (2013)
14. Saunders, M.G., Voth, G.A.: Coarse-graining methods for computational biology. Annu. Rev. Biophys. **42**, 73–93 (2013)
15. Kar, P., Feig, M.: In Biomolecular modelling and simulations. In: Karabencheva Christova, T. (ed.) vol. 96, p. 143. Elsevier Academic Press Inc., San Diego (2014)
16. Cheon, M., Chang, I., Hall, C.K.: Extending the PRIME model for protein aggregation to all 20 amino acids. Proteins **78**, 2950–2960 (2010)
17. Arkhipov, A., Yin, Y., Schulten, K.: Four-scale description of membrane sculpting by BAR domains. Biophys. J. **95**, 2806–2821 (2008)
18. Head-Gordon, T., Brown, S.: Minimalist models for protein folding and design. Curr. Opin. Struct. Biol. **13**, 160–167 (2003)
19. Matysiak, S., Clementi, C.: Minimalist protein model as a diagnostic tool for misfolding and aggregation. J. Mol. Biol. **363**, 297–308 (2006)

20. Dama, J.F., Sinitskiy, A.V., McCullagh, M., Weare, J., Roux, B., Dinner, A.R., Voth, G.A.: The theory of ultra-coarse-graining. 1. general principles. J. Chem. Theory Comput. **9**, 2466–2480 (2013)
21. Best, R.B., Hummer, G., Eaton, W.A.: Native contacts determine protein folding mechanisms in atomistic simulations. Proc. Natl. Acad. Sci. U.S.A. **110**, 17874–17879 (2013)
22. Sippl, M.J.: Knowledge-based potentials for proteins. Curr. Opin. Struct. Biol. **5**, 229–235 (1995)
23. Thompson, J.J., Tabatabaei Ghomi, H., Lill, M.A.: Application of information theory to a three-body coarse-grained representation of proteins in the PDB: insights into the structural and evolutionary roles of residues in protein structure. Proteins, **82**, 3450–3465 (2014)
24. Marrink, S.J., Risselada, H.J., Yefimov, S., Tieleman, D.P., de Vries, A.H.: The MARTINI force field: coarse grained model for biomolecular simulations. J. Phys. Chem. B **111**, 7812–7824 (2007)
25. Stark, A.C., Andrews, C.T., Elcock, A.H.: Toward optimized potential functions for protein-protein interactions in aqueous solutions: osmotic second virial coefficient calculations using the MARTINI coarse-grained force field. J. Chem. Theory Comput. **9**, 4176–4185 (2013)
26. Smith, S.W., Hall, C.K., Freeman, B.D.: Molecular dynamics for polymeric fluids using discontinuous potentials. J. Comput. Phys. **134**, 16–30 (1997)
27. Proctor, E.A., Ding, F., Dokholyan, N.V.: Discrete molecular dynamics. Wiley Interdiscip. Rev. Comput. Mol. Sci. **1**, 80–92 (2011)
28. Shirvanyants, D., Ding, F., Tsao, D., Ramachandran, S., Dokholyan, N.V.: Discrete molecular dynamics: an efficient and versatile simulation method for fine protein characterization. J. Phys. Chem. B **116**, 8375–8382 (2012)
29. Paul, G.: A complexity O(1) priority queue for event driven molecular dynamics simulations. J. Comput. Phys. **221**, 615–625 (2007)
30. Berendsen, H.J.C., Postma, J.P.M., van Gunsteren, W.F., Hermans, J.: Intermolecular Forces Reidel. Dordrecht (1981)
31. Essmann, U., Perera, L., Berkowitz, M.L., Darden, T., Lee, H., Pedersen, L.G.: A smooth particle mesh Ewald method. J. Chem. Phys. **103**, 8577–8593 (1995)
32. Hess, B.: P-LINCS: a parallel linear constraint solver for molecular simulation. J. Chem. Theory Comput. **4**, 116–122 (2008)
33. Bussi, G., Donadio, D., Parrinello, M.: Canonical sampling through velocity rescaling. J. Chem. Phys. **126**, 014101(1–7) (2007)
34. Parrinello, M., Rahman, A.: Polymorphic transitions in single crystals: a new molecular dynamics method. J App. Phys. **52**, 7182–7190 (1981)
35. Pronk, S., Páll, S., Schulz, R., Larsson, P., Bjelkmar, P., Apostolov, R., Shirts, M.R., Smith, J. C., Kasson, P.M., van der Spoel, D., Hess, B., Lindahl, E.: GROMACS 4.5: a high-throughput and highly parallel open source molecular simulation toolkit. Bioinformatics **29**, 845–854 (2013)
36. Andersen, H.C.: Molecular dynamics simulations at constant pressure and/or temperature. J. Chem. Phys. **72**, 2384–2393 (1980)
37. Martínez, L., Andrade, R., Birgin, E.G., Martínez, J.M.: PACKMOL: A package for building initial configurations for molecular dynamics simulations. J. Comput. Chem. **30**, 2157–2164 (2009)
38. Reith, D., Pütz, M., Müller-Plathe, F.: Deriving effective mesoscale potentials from atomistic simulations. J. Comput. Chem. **24**, 1624–1636 (2003)
39. Thomson, C., Lue, L., Bannerman, M.N.: Mapping continuous potentials to discrete forms. J Chem. Phys. **140**, 034105(1–9) (2014)
40. Curtis, E.M., Hall, C.K.: Molecular dynamics simulations of dppc bilayers using "LIME", a new coarse-grained model. J Phys. Chem. B **117**, 5019–5030 (2013)
41. Neal, B.L., Lenhoff, A.M.: Excluded volume contribution to the osmotic second virial coefficient for proteins. AIChE J. **41**, 1010–1014 (1995)

42. Ruppert, S., Sandler, S.I., Lenhoff, A.M.: Correlation between the osmotic second Virial coefficient and the solubility of proteins. Biotechnol. Progr. **17**, 182–187 (2001)
43. Tessier, P.M., Lenhoff, A.M., Sandler, S.I.: Rapid measurement of protein osmotic second virial coefficients by self-interaction chromatography. Biophys. J. **82**, 1620–1631 (2002)
44. Tessier, P.M., Sandler, S.I., Lenhoff, A.M.: Direct measurement of protein osmotic second virial cross coefficients by cross-interaction chromatography. Protein Sci. **13**, 1379–1390 (2004)
45. Monticelli, L., Kandasamy, S.K., Periole, X., Larson, R.G., Tieleman, D.P., Marrink, S.-J.: The MARTINI coarse-grained force field: extension to proteins. J. Chem. Theory Comput. **4**, 819–834 (2008)
46. Marin-Rimoldi, E., Shah, J.K., Maginn, E.J.: Monte Carlo simulations of water solubility in ionic liquids: a force field assessment. Fluid Phase Equilib. (2015). doi:10.1016/j.fluid.2015.07.007

Probing How Defects in Self-assembled Monolayers Affect Peptide Adsorption with Molecular Simulation

K.G. Sprenger, Yi He and Jim Pfaendtner

Abstract Due to their flexible chemical functionality and simple formulation, self-assembled monolayer (SAM) surfaces have become an ideal choice for a multitude of wide-ranging applications. However, a major issue in the preparation of SAM surfaces is naturally occurring defects that manifest in a number of different ways, including depressions in the underlying gold substrate that cause surface roughness or through incorrect self-assembly of the chains that causes domain boundary effects. Molecular simulations can provide valuable insight into the origins of these defects and the effect they have on biological and other processes. Molecular dynamics (MD) simulations have been performed on a SAM surface with a carboxylic acid/carboxylate terminal functionality and induced with two types of experimentally observed defects. The enhanced sampling method PTMetaD-WTE has been used to model the adsorption of LKα14 onto the two types of defective SAM surfaces and onto a control SAM surface with no defective chains. An advanced clustering algorithm has been used to elucidate the effect of the surface defects on the conformations of the adsorbed peptide. Results show significant structural differences arise as a result of the defects. Specifically, both types of defects lead to a near-complete loss of secondary structure of the adsorbed peptide as compared to the control simulation, in which LKα14 adopts a perfect helical structure at the SAM/water interface. On the surface with domain boundary effects, extended conformations of the peptide are stabilized, whereas on the SAM with surface roughness (i.e., chains of various lengths), random coil conformations dominate the ensemble of surface-bound structures.

Keywords Self-assembled monolayers · Surface defects · Peptide adsorption · Molecular dynamics · Enhanced sampling

---

K.G. Sprenger · J. Pfaendtner (✉)
Department of Chemical Engineering, University of Washington,
Seattle, WA 98195, USA
e-mail: jpfaendt@uw.edu

Y. He
College of Chemical and Biological Engineering, Zhejiang University,
Hangzhou 310027, Zhejiang, China

R.Q. Snurr et al. (eds.), *Foundations of Molecular Modeling and Simulation*,
Molecular Modeling and Simulation, DOI 10.1007/978-981-10-1128-3_2

1 Introduction

The formation and characterization of self-assembled monolayers (SAMs) on solid surfaces has been extensively studied for several decades. The easy preparation of SAMs with different terminal chemical functionalities has made them convenient for far-reaching and numerous applications, including bio-related technologies such as biosensors and medical implants, nano- and microfabrication, nanodevices, and corrosion protection. Experimental microscopy studies have long shown that SAMs have high concentrations of defects [1–3]; in some cases, as with the nanofabrication method of microcontact printing, naturally occurring imperfections in the SAMs were shown to play a beneficial role in the process [4]. In most cases, however, defects in the monolayers can have unexpected and perhaps undesirable consequences. Two commonly occurring defects arise from imperfections in the substrate (leading to increased surface roughness after self-assembly) and imperfections in the self-assembly process (i.e., the so-called film defects).

Though molecular simulation can offer unique insights into the consequences of SAM structural imperfections, it has only rarely been done [5–9]; limitations of small simulation cell sizes and/or insufficient sampling times have prevented the explicit exploration of defects in typical SAM modeling studies [4]. We have employed the enhanced sampling method parallel tempering metadynamics using the well-tempered ensemble (PTMetaD-WTE), which we have used successfully in several prior studies to study peptide/protein adsorption at interfaces [10–12]. A description of other simulation approaches to studying these types of problems can be found elsewhere [11].

In this work, we build on our prior simulations [11] of the model peptide LKα14 [13] adsorbing onto an ideal SAM. Past work focused on obtaining structural and thermodynamic information of adsorbed peptides, with a specific emphasis on quantitative comparison to experimental measurements of side chain orientation. However, the systems studied were very idealized due to their lack of SAM structural imperfections. In this work, we take the logical next step by studying the impact of incorporating surface defects and provide new insights into the consequences of SAM imperfections on the structure and binding thermodynamics of adsorbed biomolecules. Herein, we have performed a series of molecular dynamics (MD) simulations with PTMetaD-WTE of LKα14 adsorbing onto a carboxyl-terminated alkanethiol SAM with both substrate and film naturally occurring defects incorporated to mimic experimental observations. In addition to the simplicity of the peptide (the alpha helix organizes the side chains into a hydrophobic and charged, hydrophilic face with sequence LKKLLKLLKKLLKL), this combination of surface and peptide was chosen owing to the ease with which future experiments could be performed related to further understanding defects in SAMs. With an idealized SAM as a control, two types of defects are introduced, namely a gold depression that creates shortened alkyl chain lengths to mimic a characteristic defect in the underlying gold substrate and a characteristic film defect arising from faulty packing of the SAM (i.e., chains pointed toward and away from each other), creating domain boundary effects. We also used an advanced clustering analysis and reweighting technique to reveal large differences in surface-bound peptide conformations caused by the presence and type of incorporated SAM defect. As we discuss, this analysis is quite general and can be applied to any type of biased protein/surface simulation.

2 Methods

2.1 System Setup

System specifications are reported in Table 1, including information from a control simulation without defects from our past work [11]. Systems consisted of one LKα14 peptide, a SAM surface functionalized with a carboxylic acid/carboxylate head group, explicit TIP3P waters, and sodium ions to achieve system charge neutrality. The LKα14 peptide structure was generated with the VMD psfgen plug-in [14], and the defect-free SAM surface was based on our prior studies. LKα14 was capped with a deprotonated carboxylate group to match experimental conditions [15–23], imparting it an overall peptide charge of +5. Two types of defects were introduced into the SAM surfaces. The first type of defect mimics an experimentally observed defect in the underlying gold substrate where depressions in the gold layer lead to shortened alkyl chain lengths (hereafter referred to as a "Type I" defect, see Fig. 1).

The original surface consisted of 100 randomly alternating protonated and deprotonated chains in a 1:1 ratio to mimic a bulk pH of 7.4 [24]. Fifty chains were randomly mutated to have reduced alkyl chain lengths from 12 to 8 carbons. The same force field parameters were used for the head groups of both the healthy and mutated chains, leaving the overall surface charge of −50 unaffected. Force field parameters were taken from the AMBER99SB-ILDN force field [25] (i.e., COOH/COO from glutamic acid/glutamate). Triplicate systems were set up in this manner; distributions of the healthy/mutated chains for the 3 systems are shown in Fig. 2.

Table 1 Setup of PTMetaD-WTE simulations

Defect type	Trial	Total number of SAM chains	COO/COOH chain ratio	Mutated COO chains	Mutated COOH chains	Na^+	Waters	Box lengths (nm^3)
I	I	100	1:1	24	26	45	4334	$4 \times 5 \times 8$
I	II	100	1:1	27	23	45	4339	$4 \times 5 \times 8$
I	III	100	1:1	23	27	45	4334	$4 \times 5 \times 8$
II	N/A	70	3:4	16	24	25	4490	$4 \times 5 \times 8$
None [11]	N/A	100	1:1	0	0	45	3957	$4 \times 5 \times 8$

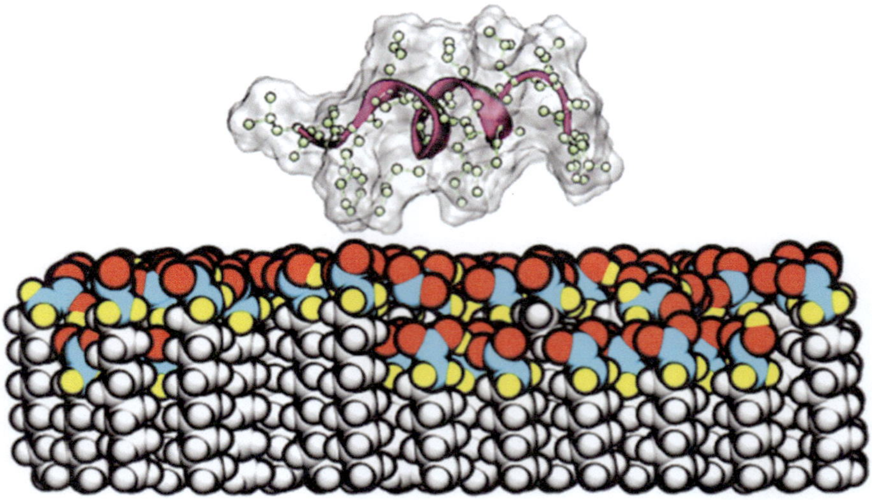

Fig. 1 Side view of LKα14 (side chains shown in space-filling representation and hydrogen not included) on a SAM surface with a Type I substrate defect causing areas of shortened alkyl chain lengths. The +*z* direction is orthogonal to the SAM surface and the +*x* direction is out of the plane of the page. Chains are *colored* to highlight frozen atoms (*silver* frozen CH_2 atoms) and head group atoms allowed to remain free during MD simulation (*teal* carbon, *yellow* hydrogen, and *red* oxygen)

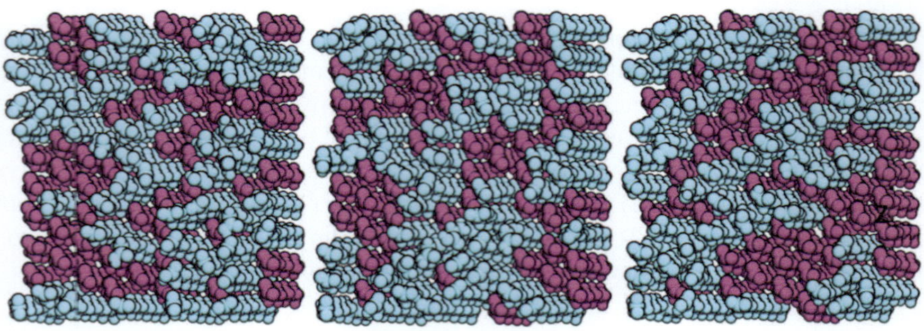

Fig. 2 Distribution of healthy to defective (i.e., short) chains for the three Type I defect simulation trials. The +z direction is out of the plane of the page. *Cyan* and *magenta* represent healthy and defective chains, respectively

The second type of defect mimics a characteristic film defect that occurs during SAM self-assembly, where alkyl chains pointing in opposite directions lead to domain boundary effects (hereafter, "Type II" defect, see Fig. 3). To introduce this defect while still maintaining the original R3 geometry and 30° normal tilt angle of the SAM chains [26], it was necessary to remove 30 of the original 100 chains. A portion of the remaining chains was then rotated about the chains' centers of

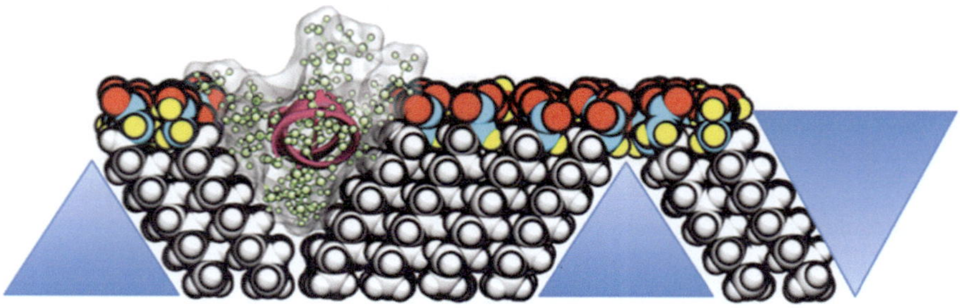

Fig. 3 Side view of LKα14 (side chains shown in space-filling representation and hydrogen not included) on a SAM surface with a Type II film self-assembly defect causing inward and outward boundary effects. The $+z$ direction is orthogonal to the SAM surface and the +y direction is out of the plane of the page. Chains are *colored* to highlight frozen atoms (*silver* frozen CH_2 atoms) and head group atoms allowed to remain free during MD simulation (*teal* carbon, *yellow* hydrogen, and *red* oxygen)

mass (minus the head groups), creating both the outward and the inward defects shown from left to right in Fig. 3. To prevent spurious interactions with the thiol group exposed at the base of the inward boundary defect, thiol groups were removed from the original surface. As all simulations used periodic boundary conditions in the x, y, and z dimensions to allow for electrostatic calculations with the particle mesh Ewald (PME) method [27], the peptide could interact with water in the triangular regions marked in blue in Fig. 3.

Simulations used the GROMACS 4.6.5 MD engine [28] with the AMBER99SB-ILDN force field [25] and the PLUMED 2.0 plug-in [29]. Box heights were chosen to permit diffusion of the peptide beyond the short-range van der Waals and Coulombic cutoff distances of 1.0 nm to experience a bulk-like state. The peptide was prevented from interacting with the image of the surface by placing a harmonic restraint on its center of mass that began acting on the peptide at a z-distance of 4.5 nm from the top of the surface. Energy minimization was performed on all surfaces with a steepest descent algorithm for 40,000 steps, followed by the minimization of the solvated peptide/surface systems where the first 6 and 10 CH_2 groups were frozen for the mutated and healthy SAM chains, respectively. Chains were frozen to prevent diffusion or melting at high temperatures and remained frozen in all ensuing simulations while movement of the head groups was unrestricted.

2.2 *System Setup*

Due to the strong binding forces that exist between the peptide and surfaces, the use of a multiscale modeling algorithm to overcome sampling challenges is essential. This type of algorithm, as applied to protein adsorption, should (1) have strong atomistic detail (e.g., be based on MD or other molecular techniques), (2) be scalable to systems of practical size, and (3) allow for quantitative comparison with experiments (e.g., in resolving the conformation and orientation of adsorbed proteins for comparison with, for example, SFG results). A method that can address all of these challenges is metadynamics (MetaD) [30, 31], which works by applying a history-dependent bias to one or more collective variables (CVs) that describe the underlying changes in a system (e.g., interfacial versus solution state structure of biomolecules in an adsorption process) in reduced dimension:

$$V(s(r),t) = W \sum_{t'=\tau_G,2\tau_G}^{t'<t} \prod_{i=1}^{N_{CV}} \exp\left[\frac{-(s_i(r) - s_i(r(t')))}{2\sigma_i^2}\right] \tag{1}$$

The added bias potential, $V(s, t)$, is added to the overall potential energy and is repulsive, Gaussian-shaped, and centered on the CV at the time of addition. This results in a net force that prevents the system from exploring previously visited states and instead encourages it to explore new regions of the CVs. To achieve smooth convergence of the bias potential, we use the well-tempered variant of metadynamics (WTM) [32]:

$$W' = \omega^* \exp\left[-\frac{V(s,t)}{k_B \Delta T}\right] \tag{2}$$

In Eq. (1), the number of CVs is given by N_{CV}, the values of which are defined by a functional mapping that relates the CV to the system's geometry, or $s(r)$. Gaussian "hills" are added every τ_G time steps with characteristic height W and width σ. WTM leads to an exponential decrease in the amount of bias added to previously explored regions of phase space (Eq. 2). The instantaneous hill height, W', is also controlled by an adjustable parameter ΔT that is related to the characteristic barrier heights in the system. In a post-processing manner, the cumulative bias from the simulation can be inverted to obtain the underlying free energy surface (FES) as projected onto the CVs [33].

Despite its capacity to greatly enhance conformational sampling, MetaD suffers from the ability of the chosen CVs to overcome hidden degrees of freedom in the system. This can be addressed with the use of parallel tempering (PT) [34, 35], which manipulates some or all degrees of freedom in a more general way (e.g., by increasing the temperature of the system). PT works by requiring many parallel simulations or "replicas" of the system that span a wide temperature range and exchange configurations periodically according to the Metropolis criterion. In this way, PT can be combined with metadynamics (PTMetaD [36]) to both increase the exploration of CV space and overcome hidden energy barriers.

The addition of sampling in the well-tempered ensemble (WTE) [37] provides an efficiency boost to the method, which has been discussed elsewhere [10]. The WTE algorithm works by using the potential energy itself as a CV and amplifying energy fluctuations (while leaving average energies of the original ensemble untouched) to increase overlap in the energy distributions of adjacent temperatures. This in turn increases the frequency of exchange between replicas and thus increases the overall efficiency of the method. The degree of amplification of the energy fluctuations is controlled via the same adjustable parameter ΔT. However, the WTE bias of the simulation is generally constructed with a different value of this parameter. Commonly, ΔT is rewritten as γ, called the bias factor, where $\gamma = (\Delta T + T)/T$ [31].

PTMetaD-WTE was used with the same procedure described in past work [11], including the use of a new functionality in PLUMED 2.0 [29] to provide a slight improvement to the method. Spanning a range of 300–450 K, 12 configurationally identical replicas were simulated in a short, 1 ns NVT PT simulation to equilibrate each replica at its respective temperature. A 10 ns WTM simulation biasing the potential energy was then performed to establish the WTE to increase sampling efficiency through increasing the spread in the system's potential energy. A bias factor of 20 was used in all WTM simulations with Gaussian hills added every ps with a width of 200 kJ/mol at an initial height of 2.0 kJ/mol.

Production runs biased two CVs for LKα14 with an additional two-dimensional MetaD bias potential. As with past work [11], the first CV biased the distance between LKα14's center of mass (COM) and the surface, whereas a second conformational CV biased the number of backbone α-helical hydrogen bond contacts. A switching function with a reference bond length of 0.25 nm was used to define the degree of the contact, which was defined between α-helical hydrogen bond donor/acceptor pairs along the peptide backbone (i.e., i, $i + 4$ pairs). The distance and conformational CVs were biased with Gaussian hill widths of 0.05 and 0.1 nm, respectively. A bias factor of 10 was used in all PTMetaD-WTE simulations with Gaussian hills added every ps at an initial bias deposition rate of 3.0 kJ/mol/ps.

3 Results and Discussion

3.1 Convergence of MetaD Simulations

To assess convergence of the PTMetaD-WTE simulations, the free energy difference between the adsorbed and solvated states was calculated as a function of time. Convergence was established when the change in the free energy difference became negligible with time. Figure 4 shows the change in the Helmholtz binding energy as a function of simulation time for each of the systems listed in Table 1. All simulations were initially run for 200 ns per replica, and all Type I defect simulations were deemed converged by the end of that time period. The Type II defect simulation was extended by 50 ns per replica to achieve convergence. Figure 4 shows that both the type of defect (i.e., Type I vs. Type II) and the distribution of the defects (i.e., Type I, trials I–III) impact the final value of the free energy change upon binding as compared to the control simulation.

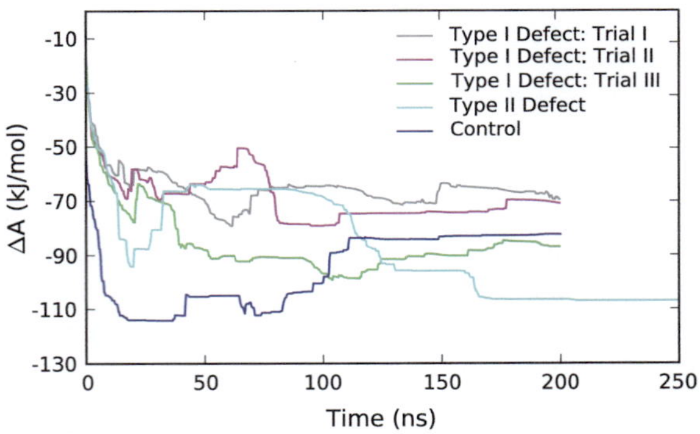

Fig. 4 Convergence of free energy differences between solvated and adsorbed states for PTMetaD-WTE simulations at 300 K. The negative value implies a decrease in free energy upon adsorption

3.2 *Clustering of Surface-Bound Structures*

Figure 5 shows the Helmholtz energy as a function of distance between LKα14 (Cα center of mass (COM)) and the surface (frozen C10 atom) for each simulation listed in Table 1. Figure 5c shows the minimum peptide/surface distance for the control simulation is approximately 1 nm; therefore, any minima in Fig. 5a, b below 1 nm represent binding to defective areas of the SAM surfaces.

To determine the effect of the defects on peptide adsorption, an RMSD-based clustering algorithm [38] was used to extract the most dominant structures in each of the wells in Fig. 5. The algorithm works by first removing external translational and rotational motions so that only the internal structural fluctuations can be characterized. A least-squares alignment between all unique pairs of structures is then performed and an RMSD value is calculated for each pair. For each structure, other structures that fall below a given cutoff value in RMSD are assigned as "neighbors". The structure with the largest number of neighbors and all of its assigned neighbors is assigned a cluster number and removed from the pool of clusters. The process is then repeated for all remaining structures until each is assigned a cluster value.

An important point should not be overlooked. The clusters obtained in the manner described above are obtained from biased MD trajectories. Therefore, it is impossible to directly compute relative cluster weights or probabilities only using the output of a clustering analysis. Instead, we employed a previously demonstrated reweighting technique [39] that makes use of the classic Torrie-Valleau umbrella sampling reweighting approach [40] with statistical weights calculated according to Eq. (3):

$$w = \exp(V_{\text{bias}}\beta) \tag{3}$$

where the bias potential in this case is obtained by using the final MetaD bias potential treated as a static biasing potential. We note for interested readers that this analysis is trivially performed within PLUMED/GROMACS by using the "-rerun" functionality of the MD engine along with the final MetaD bias (e.g., the "HILLS"

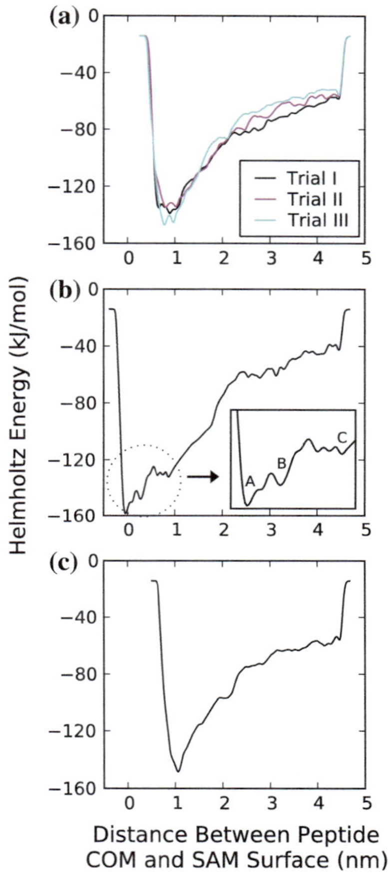

Fig. 5 Helmholtz free energy as a function of LKα14/SAM distance for PTMetaD-WTE simulations at 300 K: **a** Type I defect simulations, trials I–III; **b** Type II defect simulation, energy minima highlighted in *inset*; and **c** control simulation. Note that the relative energy scale is arbitrary owing to the trivial constant introduced in the estimation of the free energy from the MetaD bias potential

file) and the MD trajectory (i.e., in this case, the 300 K replica trajectory from the sampling scheme). Care should be taken to avoid using the portion of the trajectory that corresponds to the MetaD transient. However, in this case this is not an issue as we only clustered the 2nd half of the trajectories—far beyond the end of the transient period. With the proper statistical weights in hand for the trajectory of surface-bound structures, the final probability of each cluster is trivially calculated by normalizing and summing the individual weights (calculated via Eq. 3) for each member in each cluster.

The analysis was first performed on the trial III Type I defect simulation; since Fig. 5a shows similar free energy profiles for the three trials, we deemed analysis of a single trial to be sufficient. Skipping every second frame to reduce computation time, surface-bound structures (defined as peptide/surface distances below 1.2 nm) were clustered with an RMSD cutoff value of 0.2 nm. As noted above, we used only the second half of the trajectory for the clustering analysis to eliminate the transient part of the MetaD bias potential. Among 39,696 structures, 78 clusters

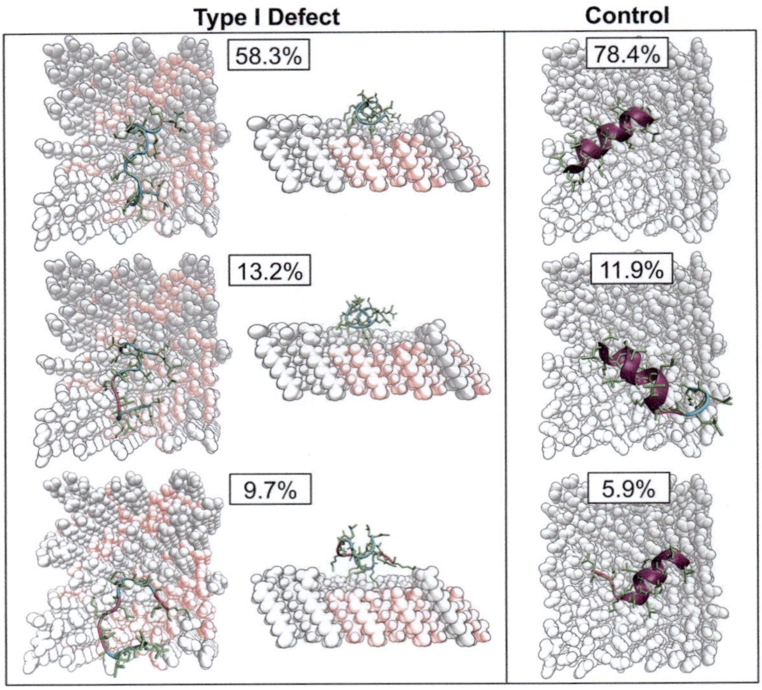

Fig. 6 Top three surface-bound cluster center conformations from a clustering analysis of the Type I, trial III defect simulation compared to the control simulation with no chain defects. Secondary structure is indicated by *peptide backbone color*: *Purple* designates an α-helix, *magenta* a turn, and *cyan* a random coil. *Silver and pink* represent healthy and defective chains, respectively

were determined. The control simulation was analyzed in a similar manner, resulting in 29 clusters from 29,848 surface-bound structures. The central conformation of each cluster, the so-called cluster centers, for the top three weighted clusters for each of these simulations, along with their respective weights, is shown in Fig. 6. Both top and side views are included for the Type I defect simulation to highlight binding to either normal or shortened alkyl chain lengths.

The first thing to note is the difference in cluster distribution between the defect and the control simulations: Conformations in the top three clusters of the defect simulation make up about 81 % of the total probability of surface-bound states, whereas conformations in the first cluster alone in the control simulation have a similar probability of existing on the surface of just over 78 %. As Fig. 6 shows, this is because areas of shortened alkyl chain lengths caused by depressions in the gold substrate below the SAM surface dramatically disrupt the helical structure that LKα14 normally adopts at interfaces, leading to a wide array of unfolded structures. Nearly, all secondary structure, indicated by the color of the peptide's backbone (i.e., magenta, cyan, and purple indicate turns, coils, and alpha helical residues, respectively), is lost with the addition of the surface defects. Unlike the central cluster conformations from the control simulation, those from the defect simulation appear to have little in common apart from a tendency toward unstructured coils, which makes sense as defective chains are randomly distributed across the surface.

The same analysis was performed on the Type II defect simulation for each of the three energy minima highlighted in Fig. 5b (i.e., A, B, and C). These minima are related to the presence of the outward boundary defect (see Fig. 3); the inward boundary defect appears to have little influence on binding. Within $\pm$ sigma of each minimum, all structures below an RMSD cutoff of 0.2 nm were clustered. This resulted in 9,885 structures in 11 clusters for minimum A, 41,203 structures in 23 clusters for minimum B, and 14,710 structures in 9 clusters for minimum C. The central cluster conformations of the clusters with the top three weights calculated for each of the minima are shown in Fig. 7.

Similar to the Type I defect results, conformations in the first cluster of energy minimum A make up about 60 % of all surface-bound states. As the distance between the peptide and the surface increases to correspond to energy minima B

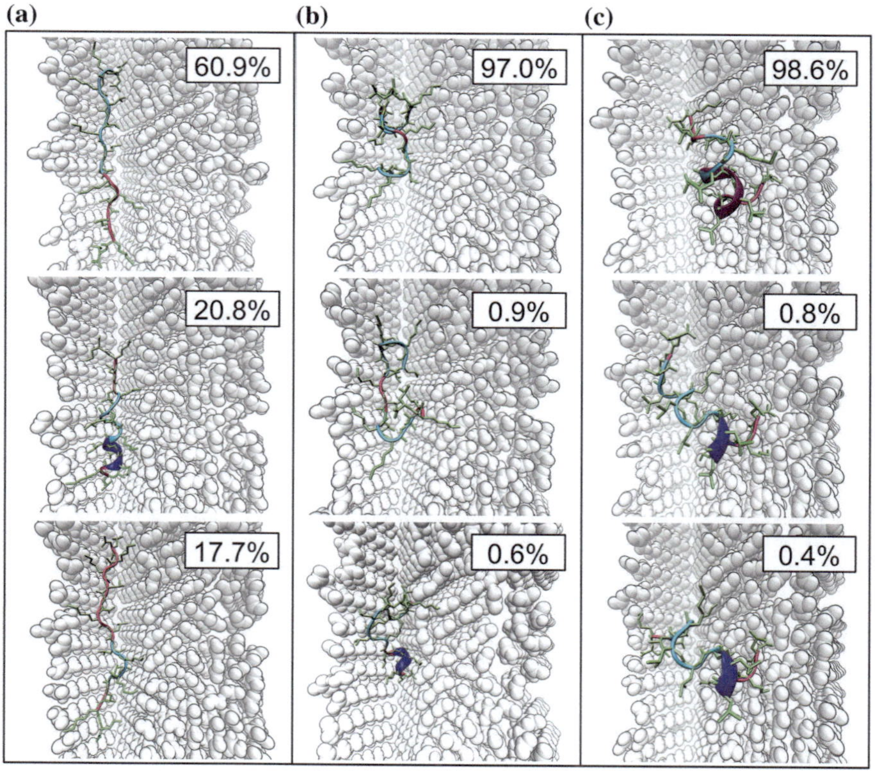

Fig. 7 Top three surface-bound cluster center conformations from a clustering analysis of the Type II defect simulation for each energy minima highlighted in Fig. 5b. Secondary structure is indicated by *peptide backbone color*: *Purple* designates an α-helix, *blue* a 3_{10}-helix, *magenta* a turn, and *cyan* a random coil

and C, however, the cluster distributions become tighter (i.e., over 95 % of all surface-bound structures reside in the top weighted cluster), similar to what was observed with the control simulation. The trends make sense given that the results for energy minimum C should most closely represent those of the control simulation due to the particular peptide/surface distance.

Deep in the hydrophobic cleft (i.e., minimum A) highly extended conformations of LKα14 are stabilized compared to structures in the control simulation, which we believe is due to the shape of the defect. Figure 5b shows binding in the pocket of minimum A is stronger than that for minimum B and much stronger than that for minimum C on top of the surface, which, as mentioned earlier, should most closely resemble the control simulation. Some α-helicity is retained on top of the surface (i.e., minimum C), as indicated by the purple color of the peptide's backbone in the cluster center conformations. However, even the mere presence of the defect causes the peptide to extend over the edge of the surface into the cavity, thereby affecting the normally helical structure of LKα14.

4 Summary/Conclusion

The enhanced sampling method PTMetaD-WTE was employed to simulate the adsorption of LKα14 to a model hydrophilic SAM with a carboxylate/carboxylic acid-terminated head group and two types of induced surface defects. Naturally occurring defects were chosen to best mimic what has been observed experimentally and included both a substrate defect and a characteristic SAM film defect. Results of free energy versus peptide/surface distance showed a difference in the location of the free energy minima for the surfaces with defects compared to a control surface with no defects. The results also indicated binding to the surface with the characteristic film defect ("Type II" defect) is much stronger than binding to the control surface, which we hypothesized is due to the specific shape of the hydrophobic cleft defect.

A clustering analysis was performed to elucidate structural differences in the bound peptide caused by the surface defects. Results showed the presence of either type of defect heavily disrupts the helical structure that LKα14 normally adopts at interfaces. In performing this analysis, peptide structures were extracted from basins, aligned, and clustered, and thus, orientation of the peptides with respect to the surface was not taken into account, only the conformation. In this case, it was not important to distinguish between orientations because charged or hydrophobic side chains dominate the surface-bound orientations. However, prior to reweighting it would be trivial to extend the clustering analysis to distinguish between orientations by subdividing further to, for example, distinguish between hydrophobic/hydrophilic patches on a peptide or protein or using other directional descriptors to account for protein orientation in conjunction with the conformational clusters.

This work will also have implications for future experimental work. Surface-guided self-assembly of proteins is growing in interest; the observed effects on peptide structure from relatively small changes in surface roughness suggest careful design of the electrostatic and van der Waals interactions at the protein/surface interface may be required. Additionally, this method could be used as a means to reverse engineer protein structure by designing and incorporating specific surface defects to control the structure of biomolecules upon adsorption.

Finally, we note that the predictions from these simulations could be directly probed with surface spectroscopies such as sum frequency generation (SFG) spectroscopy [16]. Provided self-assembly of SAMs of different chain lengths was possible, adsorption of LKα14, we predict, would reveal no appreciable SFG signal compared to neat SAMs, which reveal the expected helical structures. Likewise, using a combination of techniques such as surface plasmon resonance (SPR) and atomic force microscopy (AFM) [41], we propose it would be possible to study the expected increases in binding energy due to the film formation defects. Of course, this would depend on being able to synthesize in a controlled way the film-type defects.

Acknowledgments JP and KGS acknowledge financial support from NSF award CBET-1264459. This work was facilitated through the use of computational, storage, and networking infrastructure provided by the Hyak supercomputer system, supported in part by the University of Washington and the UW Student Technology Fee Proposal program (award 2015-028). This research was also supported by the National Natural Science Foundation of China through grant numbers 21450110411, 21476191, and 91434110, and by the Scientific Research Fund of the Zhejiang Provincial Education Department through grant number Y201329422.

References

1. McDermott, C.A., McDermott, M.T., Green, J.-B., Porter, M.D.: Structural origins of the surface depressions at alkanethiolate monolayers on Au(111): a scanning tunneling and atomic force microscopic investigation. J. Phys. Chem. **99**, 13257–13267 (1995)
2. Noh, J., Hara, M.: Molecular-scale desorption processes and the alternating missing-row phase of alkanethiol self-assembled monolayers on Au(111). Langmuir **17**, 7280–7285 (2001)
3. Godin, M., Williams, P.J., Tabard-Cossa, V., Laroche, O., Beaulieu, L.Y., Lennox, R.B., Grütter, P.: Surface stress, kinetics, and Structure of alkanethiol self-assembled monolayers. Langmuir **20**, 7090–7096 (2004)
4. Gannon, G., Greer, J.C., Larsson, J.A., Thompson, D.: Molecular dynamics study of naturally occurring defects in self-assembled monolayer formation. ACS Nano **4**, 921–932 (2010)
5. Vemparala, S., Karki, B.B., Kalia, R.K., Nakano, A., Vashishta, P.: Large-scale molecular dynamics simulations of alkanethiol self-assembled monolayers. J. Chem. Phys. **121**, 4323–4330 (2004)
6. Prathima, N., Harini, M., Rai, N., Chandrashekara, R.H., Ayappa, K.G., Sampath, S., Biswas, S.K.: thermal study of accumulation of conformational disorders in the self-assembled monolayers of C_8 and C_{18} alkanethiols on the Au(111) surface. Langmuir **21**, 2364–2374 (2005)
7. Jiang, L., Sangeeth, C.S.S., Yuan, L., Thompson, D., Nijhuis, C.A.: One-nanometer thin monolayers remove the deleterious effect of substrate defects in molecular tunnel junctions. Nano Lett. (2015)

8. O'Mahony, S., O'Dwyer, C., Nijhuis, C.A., Greer, J.C., Quinn, A.J., Thompson, D.: Nanoscale dynamics and protein adhesivity of alkylamine self-assembled monolayers on graphene. Langmuir **29**, 7271–7282 (2013)
9. Ahn, Y., Saha, J.K., Schatz, G.C., Jang, J.: Molecular dynamics study of the formation of a self-assembled monolayer on gold. J. Phys. Chem. C **115**, 10668–10674 (2011)
10. Deighan, M., Bonomi, M., Pfaendtner, J.: Efficient simulation of explicitly solvated proteins in the well-tempered ensemble. JCTC **8**, 2189–2192 (2012)
11. Deighan, M., Pfaendtner, J.: Exhaustively sampling peptide adsorption with metadynamics. Langmuir **29**, 7999–8009 (2013)
12. Levine, Z.A., Fischer, S.A., Shea, J.-E., Pfaendtner, J.: Trp-Cage folding on organic surfaces. J. Phys. Chem. B. **119**, 10417–10425 (2015)
13. DeGrado, W.F., Lear, J.D.: Induction of peptide conformation at apolar water interfaces. 1. a study with model peptides of defined hydrophobic periodicity. J. Am. Chem. Soc. **107**, 7684–7689 (1985)
14. Humphrey, W., Dalke, A., Schulten, K.: VMD: visual molecular dynamics. J. Mol. Graphics. **14**, 33–38 (1996)
15. Weidner, T., Samuel, N.T., McCrea, K., Gamble, L.J., Ward, R.S., Castner, D.G.: Assembly and structure of α-helical peptide films on hydrophobic fluorocarbon surfaces. Biointerphases **5**, 9–16 (2010)
16. Weidner, T., Apte, J.S., Gamble, L.J., Castner, D.G.: Probing the orientation and conformation of α-helix and β-strand model peptides on self-assembled monolayers using sum frequency generation and nexafs spectroscopy. Langmuir **26**, 3433–3440 (2010)
17. Mermut, O., Phillips, D.C., York, R.L., McCrea, K.R., Ward, R.S., Somorjai, Ga: In situ adsorption studies of a 14-amino acid leucine-lysine peptide onto hydrophobic polystyrene and hydrophilic silica surfaces using quartz crystal microbalance, atomic force microscopy, and sum frequency generation vibrational spectroscopy. J. Am. Chem. Soc. **128**, 3598–3607 (2006)
18. York, R.L., Browne, W.K., Geissler, P.L., Somorjai, G.A.: Peptides adsorbed on hydrophobic surfaces—a sum frequency generation vibrational spectroscopy and modeling study. Isr. J. Chem. **47**, 51–58 (2007)
19. York, R.L., Mermut, O., Phillips, D.C., McCrea, K.R., Ward, R.S., Somorjai, G.A.: Influence of ionic strength on the adsorption of a model peptide on hydrophilic silica and hydrophobic polystyrene surfaces: insight from SFG vibrational spectroscopy. J. Phys. Chem. C **111**, 8866–8871 (2007)
20. Apte, J.S., Gamble, L.J., Castner, D.G., Campbell, C.T.: Kinetics of leucine-lysine peptide adsorption and desorption at -CH3 and -COOH terminated alkylthiolate monolayers. Biointerphases **5**, 97–104 (2010)
21. Long, J.R., Oyler, N., Drobny, G.P., Stayton, P.S.: Assembly of α-helical peptide coatings on hydrophobic surfaces. J. Am. Chem. Soc. **124**, 6297–6303 (2002)
22. Phillips, D.C., York, R.L., Mermut, O., McCrea, K.R., Ward, R.S., Somorjai, G.A.: Side chain, chain length, and sequence effects on amphiphilic peptide adsorption at hydrophobic and hydrophilic surfaces studied by sum-frequency generation vibrational spectroscopy and quartz crystal microbalance. J. Phys. Chem. C **111**, 255–261 (2007)
23. Apte, J.S., Collier, G., Latour, R.A., Gamble, L.J., Castner, D.G.: XPS and ToF-SIMS investigation of α-helical and β-strand peptide adsorption onto SAMs. Langmuir **26**, 3423–3432 (2010)
24. Fears, K.P., Creager, S.E., Latour, R.A.: Determination of the surface pK of carboxylic- and amine-terminated alkanethiols using surface plasmon resonance spectroscopy. Langmuir **24**, 837–843 (2008)
25. Lindorff-Larsen, K., Piana, S., Palmo, K., Maragakis, P., Klepeis, J.L., Dror, R.O., Shaw, D. E.: improved side-chain torsion potentials for the amber ff99SB protein force field. Proteins **78**, 1950–1958 (2010)
26. Ulman, A., Eilers, J.E., Tillman, N.: Packing and molecular orientation of alkanethiol monolayers on gold surfaces. Langmuir **5**, 1147–1152 (1989)

27. Essmann, U., Perera, L., Berkowitz, M.L., Darden, T., Lee, H., Pedersen, L.G.: A smooth particle mesh ewald method. J. Chem. Phys. **103**, 8577–8593 (1995)
28. Hess, B., Kutzner, C., van der Spoel, D., Lindahl, E.: GROMACS 4: algorithms for highly efficient, load-balanced, and scalable molecular simulation. JCTC. **4**, 435–447 (2008)
29. Tribello, G.A., Bonomi, M., Branduardi, D., Camilloni, C., Bussi, G.: PLUMED 2: new feathers for an old bird. Comput. Phys. Commun. **185**, 604–613 (2014)
30. Laio, A., Parrinello, M.: Escaping free-energy minima. PNAS **99**, 12562–12566 (2002)
31. Barducci, A., Pfaendtner, J., Bonomi, M.: Tackling sampling challenges in biomolecular simulations. Methods Mol. Bio. **1215**, 151–171 (2015)
32. Barducci, A., Bussi, G., Parrinello, M.: Well-tempered metadynamics: a smoothly converging and tunable free-energy method. Phys. Rev. Lett. **100**, 020603 (2008)
33. Dama, J.F., Parrinello, M., Voth, G.A.: Well-tempered metadynamics converges asymptotically. Phys. Rev. Lett. **112**, 240602 (2014)
34. Hansmann, U.H.E.: Parallel tempering algorithm for conformational studies of biological molecules. Chem. Phys. Lett. **281**, 140–150 (1997)
35. Sugita, Y., Okamoto, Y.: Replica-exchange molecular dynamics method for protein folding. Chem. Phys. Lett. **314**, 141–151 (1999)
36. Bussi, G., Gervasio, F., Laio, A., Parrinello, M.: Free-energy landscape for beta hairpin folding from combined parallel tempering and metadynamics. J. Am. Chem. Soc. **128**, 13435 (2006)
37. Bonomi, M., Parrinello, M.: Enhanced sampling in the well-tempered ensemble. Phys. Rev. Lett. **104**, 190601 (2010)
38. Daura, X., Gademann, K., Jaun, B., Seebach, D., van Gunsteren, W.F., Mark, A.E.: peptide folding: when simulation meets experiment. Angew. Chem. Int. Ed. **38**, 236–240 (1999)
39. Branduardi, D., Bussi, G., Parrinello, M.: Metadynamics with adaptive Gaussians. JCTC **8**, 2247–2254 (2012)
40. Torrie, G.M., Valleau, J.P.: Nonphysical sampling distributions in Monte Carlo free-energy estimation: umbrella sampling. J. Comput. Phys. **23**, 187–199 (1977)
41. Thyparambil, A.A., Wei, Y., Latour, R.A.: Determination of peptide-surface adsorption free energy for material surfaces not conducive to SPR or QCM using AFM. Langmuir **28**, 5687–5694 (2012)

Development of a Coarse-Grained Water Forcefield via Multistate Iterative Boltzmann Inversion

Timothy C. Moore, Christopher R. Iacovella and Clare McCabe

Abstract A coarse-grained water model is developed using multistate iterative Boltzmann inversion. Following previous work, the *k*-means algorithm is used to dynamically map multiple water molecules to a single coarse-grained bead, allowing the use of structure-based coarse-graining methods. The model is derived to match the bulk and interfacial properties of liquid water and improves upon previous work that used single state iterative Boltzmann inversion. The model accurately reproduces the density and structural correlations of water at 305 K and 1.0 atm, stability of a liquid droplet at 305 K, and shows little tendency to crystallize at physiological conditions. This work also illustrates several advantages of using multistate iterative Boltzmann inversion for deriving generally applicable coarse-grained forcefields.

Keywords Interface · Pressure · Crystallization · Surface tension

1 Introduction

Coarse-grained (CG) models have proven to be useful in many fields of chemical research [1–10], allowing molecular simulations to be performed on larger system sizes and access longer timescales than is possible with atomistic-level models,

---

T.C. Moore · C.R. Iacovella (✉) · C. McCabe (✉)
Department of Chemical and Biomolecular Engineering,
Vanderbilt University, Nashville, TN 37235, USA
e-mail: christopher.r.iacovella@Vanderbilt.Edu

C. McCabe
e-mail: c.mccabe@Vanderbilt.Edu

T.C. Moore · C.R. Iacovella · C. McCabe
Vanderbilt University Center for Multiscale Modeling
and Simulation (MuMS), Nashville, TN 37235, USA

C. McCabe
Department of Chemistry, Vanderbilt University, Nashville, TN 37235, USA

R.Q. Snurr et al. (eds.), *Foundations of Molecular Modeling and Simulation*,
Molecular Modeling and Simulation, DOI 10.1007/978-981-10-1128-3_3

enabling complex phenomena such as hierarchical self-assembly to be described [11, 12]. In CG simulations of aqueous systems, especially ones with significant amounts of hydrophobic and/or hydrophilic interactions, the water model is important and can have a major impact on the resulting properties of the system [13].

While the assignment of atoms to CG beads (i.e., defining the CG mapping) is relatively straightforward for most chemical systems (e.g., aggregating four methyl groups bonded in sequence into a single CG bead), mapping an atomistic water trajectory to the CG level (i.e., grouping several water molecules into a single CG bead) is not as well-defined given the lack of permanent bonds between water molecules. Even if a mapping were chosen, water molecules will diffuse away from their initial clusters over time, such that the initial mapping is no longer representative of the local clustering of water. This ambiguity presents a problem for structure-based methods that require an atomistic configuration to be mapped to the corresponding CG configuration, e.g., to generate a target radial distribution function (RDF) against which the forcefield is optimized. As such, the majority of many-to-one CG models of water (i.e., where one CG bead represents multiple water molecules) have instead been derived by assuming a functional form of the forcefield and optimizing the associated parameters to match selected physical properties of water, such as density, vaporization enthalpy, surface tension, etc. [13–19]. For example, Chiu et al. developed a 4:1 CG water forcefield by optimizing the parameters of a Morse potential to accurately reproduce the surface tension and density of liquid water [18]. Despite capturing the interfacial properties and density, this potential overestimates structural correlations, as one might expect given that structural data was not used in its optimization.

Recently, Hadley and McCabe [20] proposed a method for mapping configurations of atomistic water to their CG representations using the *k*-means clustering algorithm. Subsequently in related work, van Hoof et al. [21] developed the CUMULUS method for mapping atoms to CG beads. Both methods enable dynamic mapping of multiple water molecules to a single CG bead, allowing structure-based schemes to be used. Here, dynamic refers to a CG mapping that changes over the course of the atomistic trajectory, i.e., different water molecules are assigned to different CG beads in each frame of the atomistic trajectory. Both works employed the iterative Boltzmann inversion (IBI) [22] method to derive the intermolecular interaction by optimizing a numerical, rather than analytical, potential to reproduce RDFs calculated from the atomistic-to-CG mapped configurations [20, 21]. The forcefields derived are similar and show good agreement with the structural properties and density of the atomistic water models studied. However, neither model is able to accurately reproduce interfacial properties, since they were derived solely from bulk fluid data. This failure to capture interfacial properties is a consequence of the single-state nature of the IBI approach and may alter the balance of hydrophobic and hydrophilic interactions when using these water models in multicomponent systems.

Recently, the multistate IBI (MS IBI) method [23] was developed as an extension of the original IBI approach, with the goal of reducing state dependence and structural artifacts often found in IBI-based potentials [24–26]. While IBI-based potentials have been derived that show some degree of transferability [26–28] a significant issue related to the IBI method is that a multitude of potentials can give rise to similar RDFs, and the method cannot necessarily differentiate which of the many potentials is most accurate, as only RDF matching is considered. MS IBI operates based on the idea that different thermodynamic states will occupy different regions of potential "phase space" (i.e., regions where potentials give rise to similar RDFs), and that the most transferable, and thus most accurate, potential lies in the overlap of phase space for the different states. That is, by optimizing a potential simultaneously against multiple thermodynamic states, MS IBI provides constraints to the optimization, forcing the method to derive potentials that exist in this overlap region, and thus are transferable among the states considered. The MS IBI approach has been shown to reduce state dependence and improve the quality of the derived potentials, as compared to the original IBI method [23].

In this work, multistate iterative Boltzmann inversion (MS IBI) is used to derive an intermolecular potential that captures both bulk and interfacial properties of water, improving upon the CG water model of Hadley and McCabe [20]. Again, optimizations are carried out using the MS IBI method, where both bulk and interfacial systems are used simultaneously as target conditions for the optimization. MS IBI is also used, for the first time, in a multi-ensemble context, enabling optimizations in both the canonical (NVT) and isothermal-isobaric (NPT) ensembles to be performed simultaneously to derive the density-pressure relationship of the system. To further constrain the optimization, a slightly modified version of the Chiu et al. CG water forcefield, originally optimized for surface tension, is used as a starting condition, allowing the MS IBI method to make specific modifications to the potential to improve structural properties. The remainder of the paper is organized as follows: In Methods, a brief overview of the *k*-means clustering and MS IBI algorithms is given and the models used are described. The potential derivation is then presented, validated, and compared to existing CG water models in the Results section and finally, conclusions are drawn about the applicability of the derived CG model and the broader applicability of the MS IBI method discussed.

2 Methods

2.1 k-Means Clustering Algorithm

Mapping a water trajectory to a many-to-one CG level is inherently different than mapping a larger molecule's trajectory, since for water, atoms mapped into a single CG bead necessarily exist on different molecules. Furthermore, the water molecules mapped to a common bead are not likely to remain associated throughout the full simulation because of thermal diffusion. A dynamic mapping scheme is therefore required to generate CG structures from atomistic configurations for water. Following the work of Hadley and McCabe [20], the k-means algorithm has been used to map atomistic water trajectories to the CG level. In short, k-means is a clustering algorithm that is used to find clusters of data points in a large data set. The positions of the water molecules are here analogous to the points in the data set and waters mapped to a single bead are analogous to the clusters. Additional details on the algorithm can be found elsewhere [20, 29]. While the k-means algorithm can be used to group together any number of water molecules, a 4:1 mapping is chosen, as this was found in prior work to provide the best balance between accuracy and computational efficiency [20] and 4:1 models are common in the literature [17, 18, 20].

2.2 Multistate Iterative Boltzmann Inversion Method

MS IBI was used to derive the intermolecular potential between water beads. The goal of MS IBI is to derive a single potential that can be used over a range of thermodynamic states. As an extension of the original IBI method [22], the potential is updated based on the average differences in CG and target RDFs at multiple states (i.e., a single potential for each pair is updated based on RDFs from multiple states). The potential is adjusted according to

$$V_{i+1}(r) = V_i(r) - \frac{1}{N}\sum_s \alpha_s(r) k_B T_s \ln\left[\frac{g_s^*(r)}{g_s^i(r)}\right], \tag{1}$$

where $V_i(r)$ is the pair potential as a function of separation r at the ith iteration; N the number of target states; $\alpha_s(r)$ an effective weighting factor for state s, allowing more or less emphasis to be put on a particular target state; k_B the Boltzmann constant, T_s the absolute temperature of state s; $g_s^i(r)$ the RDF from the CG simulation at state s using $V_i(r)$; and $g_s^*(r)$ the target RDF from state s. $\alpha_s(r)$ was chosen to be a linear function of the form

$$\alpha_s(r) = \alpha_{0,s}\left(1 - \frac{r}{r_{cut}}\right), \tag{2}$$

such that $\alpha_s(r_{cut}) = 0$ and the potential remains 0 for $r \geq r_{cut}$. This form of $\alpha_s(r)$ also places more emphasis on the short-ranged part of the potential to suppress long-range structural artifacts.

An initial potential is assumed for each pair interaction. In theory, there are no restrictions on the initial potential, so it may take any form; however, in practice, the initial potential is often taken to be the potential of mean force (PMF) calculated from the Boltzmann inverted RDF. In this work, rather than taking an average of the PMFs over the states used, the initial potential used was chosen to be a slightly modified version of Chiu et al.'s water model, as discussed below. That is, rather than starting from an initial potential that is likely to do a poor job of predicting the behavior, we start from a robust starting point as the Chiu et al. potential is known to accurately reproduce several properties of water.

A CG simulation is then run with the initial potential. Based on the RDFs from the CG simulation, the potential is updated according to Eq. (1). The updated potential is used as input to the next cycle, and the process is repeated until some stopping criterion is met. Here, the stopping criterion is determined using the following fitness function

$$f_{\text{fit}} = 1 - \frac{\int_0^{r_{\text{cut}}} \mathrm{d}r \left| g^i(r) - g^*(r) \right|}{\int_0^{r_{\text{cut}}} \mathrm{d}r \left| g^i(r) \right| + \left| g^*(r) \right|}, \quad (3)$$

where the optimization is stopped when the value of f_{fit} exceeds a specified value (i.e., meets some tolerance), given below.

2.3 *Models*

Atomistic simulations of pure water were performed with the TIP3P model [30]. All atomistic systems contained 5,832 water molecules and were simulated in LAMMPS [31, 32] using a 1 fs timestep. A cutoff distance of 12 Å was used for the van der Waals interactions; long-range electrostatics were handled with the PPPM method with a 12 Å real space cutoff. Three distinct states were simulated: bulk, NVT at 1.0 g/mL and 305 K; bulk, NPT at 305 K and 1.0 atm; and an NVT droplet state at 305 K, where the box from the bulk NVT state was expanded by a factor of 3 in one direction. Each atomistic simulation was run for 7 ns. The atomistic trajectories were mapped to the CG level using the *k*-means algorithm. Target RDFs were calculated from the final 5 ns of the mapped trajectory from each state (bulk NVT, bulk NPT, and droplet NVT). MS IBI was performed using the target data from each of the three states. The initial guess of the potential is given as a Morse potential of the form

$$V(r) = D_e \left(e^{-2\beta\left(r - r_{\text{eq}}\right)} - 2e^{-\beta\left(r - r_{\text{eq}}\right)} \right), \quad (4)$$

where r_{eq} is the location of the potential minimum, $-D_e$ is the value of the potential minimum, and β is related to the width of the potential well. Parameters are taken to be those from Chiu et al: $D_e = 0.813$ kcal/mol, $\beta = 0.556$ Å^{-1}, and $r_{\text{eq}} = 6.29$ Å, however, we note that the potential was adjusted so that $\beta = 0.5$ Å^{-1} for $r < r_{\text{eq}}$. This change was made to increase sampling at small separations, because numerical issues arise in the potential update when the CG RDF is zero but the target RDF is nonzero. This modification of the potential will slightly alter the properties as compared to the original model, as discussed below. The potential update scaling factor $\alpha_{0,s}$ (see Eqs. 1 and 2) was set to 0.7 to avoid large updates to the potential. The optimizations were stopped when $f_{\mathrm{fit}} \geq 0.98$ and $f_{\mathrm{fit}}(i) - f_{\mathrm{fit}}(i-1) < 0.001$ for each state.

All optimizations were performed with the open-source MS IBI Python package we developed [33], which calls HOOMD-Blue [34–36] to run the CG simulations and uses MDTraj [37, 38] for RDF calculations and file-handling. CG simulations were run at the same states as the atomistic systems. Initial CG configurations were generated from the CG-mapped atomistic trajectories at each state. As a result of the 4:1 mapping, CG water simulations contained 1,458 water beads. All CG simulations were run with a 10 fs timestep. The derived CG potential was set to 0 beyond the cutoff of 12 Å.

The surface tension γ of the droplet state was calculated as

$$\gamma = \frac{1}{2} L_z \left\langle P_{zz} - \frac{P_{xx} + P_{yy}}{2} \right\rangle,$$

where L_z is the length of the box in the expanded direction, P_{zz} is the pressure component in the direction normal to the liquid-vapor interfaces, P_{xx} and P_{yy} are the pressure components in the directions lateral to the interfaces, and the angle brackets denote a time average. The factor of ½ is included to account for the two interfaces that are present in the droplet simulation setup.

3 Results and Discussion

3.1 Modified Chiu Potential

Since the MS IBI optimization of water uses a modified version of the Chiu, et al. potential as an initial guess, we first consider the impact of modifying the potential to create a softer repulsion. Figure 1 plots the RDFs of the three target states for the original and modified potentials and the RDF of the 4:1 mapped state (i.e., the target data used later for the MS IBI optimization). The peak location of the NVT state is relatively unchanged; however, upon modification, there is a slight shift in the first peak for the NPT and interfacial states, allowing the model to access smaller separations, as was intended and required for the potential update scheme. The softer potential allows closer contact and thus allows the MS IBI algorithm to modify this region of the potential where the 4:1 mapped atomistic water has non-zero values of the RDF. The density predicted with both potentials is the same (0.991 ± 0.003 g/mL); however, due to softening the potential, the calculated surface tension of the droplet changes from 70.3 to 45 mN/m after the modification, although this value is still sufficient for the droplet to maintain a stable interface.

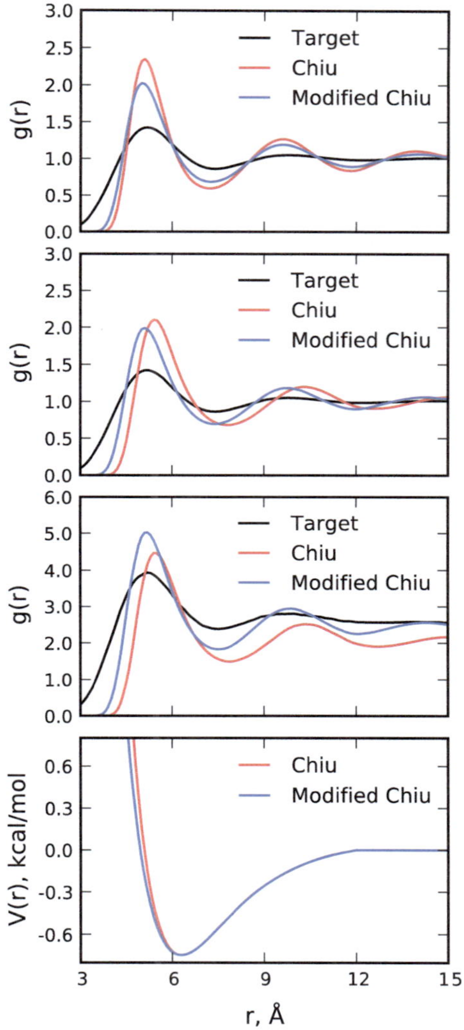

Fig. 1 RDFs from simulations using the original and modified Chiu potentials. *Top* NVT; *top-middle* NPT; *bottom-middle* interface; *bottom* comparison of the two potentials

These surface tension values agree favorably with that of TIP3P water, which is reported to have a surface tension of 52.3 mN/m at 300 K [39].

3.2 *Potential Derivation and Validation*

Starting from the modified Morse potential of Chiu, et al., the new water forcefield is optimized using the bulk NVT and NPT states and the interfacial state. This potential is chosen as the initial starting guess, rather than an arbitrary starting point, as the unmodified version has been shown to accurately reproduce many properties of water (e.g., density and surface tension), but overestimates the structural correlations. The use of MS IBI should allow for modification of this potential, such that it is able to reproduce structural quantities. The results of the potential derivation are summarized in Fig. 2, where it is clear that the modified Chiu, et al. potential (i.e., step 0) overestimates the structural correlations, as was also seen in Fig. 1 for both the modified and original potentials. After only a few iterations, the RDFs match the targets with a high degree of accuracy. This trend is shown in Fig. 3, which plots the fitness value from Eq. (3) as a function of iteration. The value of f_{fit} changes most rapidly in the first 3 steps of the optimization. After 10 iterations, the stopping criteria are met and the optimization stopped. While the

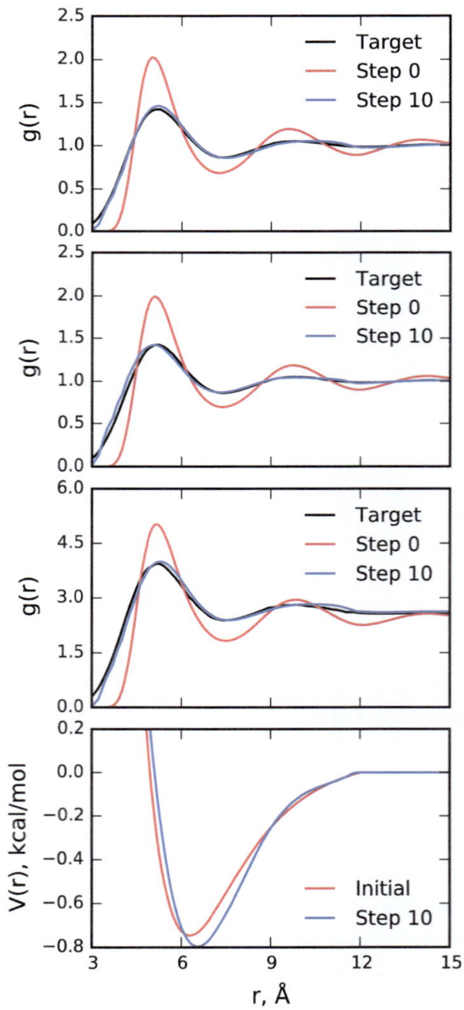

Fig. 2 RDFs and potentials from the MS IBI potential derivation. *Top* NVT; *middle-top* NPT; *middle-bottom* interface; *bottom* potentials. The initial potential shows significant structural correlations missing from the target data. The derived potential at ten iterations shows excellent structural agreement with the target

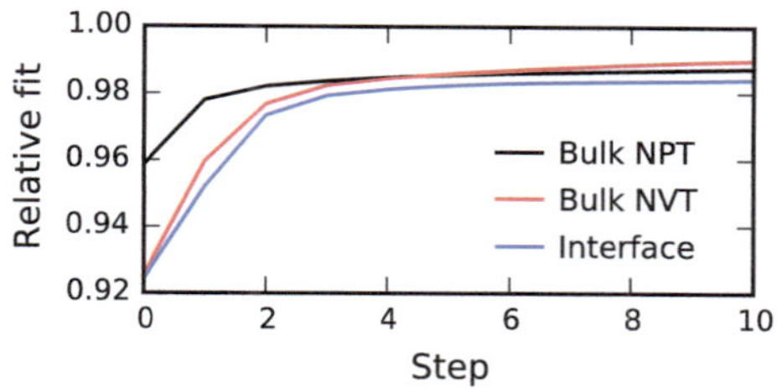

Fig. 3 f_{fit} from Eq. (3) as a function of iteration in the potential derivation. Convergence with the criterion is found after 10 iterations

changes to the potential are small, there is a noticeable shift in the location of the minimum to a slightly larger r value and the potential becomes slightly more attractive. Although the shape of the attractive well is mostly unchanged, the potential more rapidly decays to 0 than the original Morse potential at larger r values, while the shape of the repulsive regime at small r is changed slightly. These subtle changes to the potential are sufficient to create significant changes in the RDF and provide excellent convergence of the structural correlations. These changes are made possible by modifying a numerical potential rather than adjusting parameters for an analytical potential. Note that in Figs. 1 and 2 the RDFs from the interfacial state do not decay to 1 at large r. This is due to the fact that 2/3 of the box is essentially devoid of particles, but the RDF is normalized based on the volume of the whole simulation box. This has no effect on the potential update scheme, as both the target and CG RDFs are normalized by the same factor, which cancels out in Eq. (1).

In addition to accurately capturing the RDFs, the multi-ensemble approach provides an accurate estimate of the density at 305 K and 1 atm. NPT simulations performed using the optimized CG forcefield find a density of 1.027 ± 0.006 g/mL, compared to 1.037 ± 0.004 g/mL for TIP3P water which was used to generate target data. This approach is successful because the RDFs will not match if the pressure-density relationship is not satisfied, as the density is implicitly represented in Eq. (1) through the RDF terms (i.e., the RDFs at the NPT state will not match the target RDFs if the density is significantly different than the density of the target state). In contrast, the original IBI method proposed the use of a pressure correction term of the form $\Delta V(r) = A(1 - r/r_{cut})$ to account for the pressure [22]. This approach has been successful, but requires a somewhat arbitrary estimate of the parameter A. While a method exists for estimating A based on the virial expression [40], some degree of trial-and-error is still necessary. Furthermore, the multi-ensemble approach within MS IBI does not require direct calculation of the pressure, which often demonstrates considerable fluctuations, providing a simpler route to account for pressure in the CG model.

Calculation of the surface tension of the derived MS IBI potential yields a value of 42 mN/m, lower than the original Chiu, et al. potential (70.3 mN/m) which was optimized to match experiment, but only slightly perturbed from the modified potential (45 mN/m). This reduction in surface tension appears directly related to the softening of the potential, although, we note that this softening is required to provide an accurate match of the structure and that this value reasonably approximates the surface tension of the atomistic TIP3P model used as target data (52.3 mN/m at 300 K) [39].

3.3 Validation and Comparison to Other Models

To further explore the efficacy of the MS IBI-derived model, comparisons are made to other CG water models in the literature, namely, the *k*-means based potential of Hadley and McCabe [20] derived via the single state (SS) IBI procedure (here referred to as the SS IBI potential) and the MARTINI potential [17]. These models were chosen because they are short-ranged, non-polarizable, and 4:1 models. For reference, these potentials are plotted in Fig. 4. Note that the MS IBI and SS IBI potentials are numerical (as they were derived via IBI), while the MARTINI potential is represented by a 12-6 Lennard-Jones potential with a well depth of 1.195 kcal/mol located at a separation of 5.276 Å. Note that all of the potentials considered in this paper provide a close estimate of the density of water at 1 atm and 305 K, as reported in Table 1.

First considering the SS IBI potential, it can be seen that the well depth is approximately 0.5 kcal/mol weaker than the MS IBI potential and shifted to larger separations. While this has little impact on the density or the structural correlations of the bulk states (not shown), simulations of droplets show that the interfacial properties are not sufficiently captured. Specifically, as shown in Fig. 5, simulations of atomistic TIP3P, SS IBI, and MS IBI water were performed with interfaces. From these it can be clearly seen that the SS IBI potential model fills the box, rather

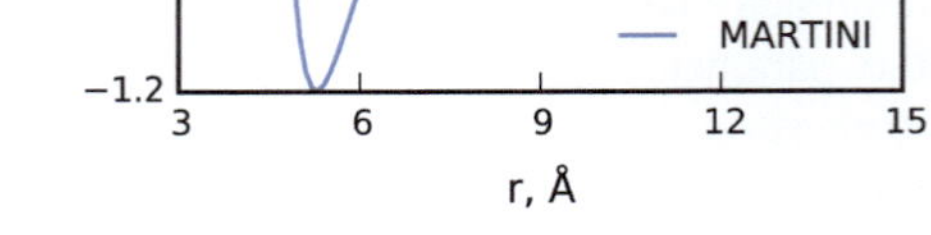

Fig. 4 Interaction potentials from the CG water models compared in this work. The MS IBI and SS IBI potentials are numerical, derived with structure-based methods. MARTINI is a Lennard-Jones 12-6 potential

Table 1 Density of water at 305 K, 1 atm calculated with different models

Model	Density (g/mL)
TIP3P	1.037 ± 0.004
MS IBI	1.027 ± 0.006
SS IBI	1.083 ± 0.008
MARTINI	1.015 ± 0.003
Chiu	0.991 ± 0.003

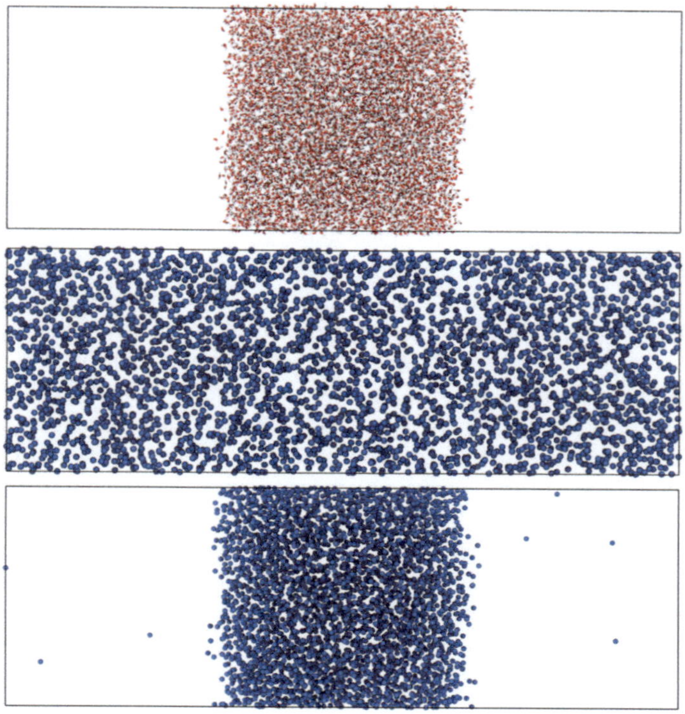

Fig. 5 Simulation snapshots of droplets using the various models discussed. *Top* all-atom TIP3P; *middle* SS IBI; *bottom* MS IBI. Atomistic and MS IBI models agree, producing a system with a stable interface, whereas SS IBI does not form a stable interface

than maintaining an interface. In contrast, the MS IBI model maintains a stable interface in agreement with the atomistic model. Thus, while an exact match to the experimental surface tension is not found for the MS IBI potential, as discussed above, it is still sufficiently strong to maintain a clear interface, providing a significant improvement over the SS IBI potential. We note that the difference between the SS IBI and MS IBI potentials is likely related to the aforementioned issue whereby many potentials can give rise to matching RDFs, and SS IBI provides no means to determine which ones are most physical. This limitation is overcome by the use of the interfacial state during the MS IBI optimization.

It is also important that the potential is not so strong that the system can solidify at physiological conditions. For example, the MARTINI water model is known to spontaneously crystallize at physiologically relevant temperatures [17]. This phenomenon is enhanced by the presence of interfaces (e.g., a lipid bilayer surface), and requires the addition of unphysical "antifreeze" particles to avoid crystallization. While we note that modifications to the MARTINI water model exist (e.g., adding charge polarization) [41, 42], only the original MARTINI model was tested, since it more closely resembles the model derived via MS IBI (i.e., both represent 4 water molecules as a single, spherically symmetric interaction site). To test the crystallization tendency, a nucleation site is generated with the following protocol. A crystalline state is generated by running a simulation with the MS IBI potential in the NVT ensemble. During this simulation, the temperature is decreased from 305 to 1 K over 10 ns. A subsequent CG simulation is run at 1000 K, where the middle-most 1/8th of the beads are kept fixed, resulting in a configuration that contains a crystal seed surrounded by a fluid of CG water beads. The beads in the crystal seed are kept fixed in the nucleation site simulations, with interactions identical to the fluid interactions. While neither model shows a tendency to freeze at 305 K in the absence of a nucleation site over a 100 ns simulation, the MARTINI model rapidly crystallizes in the presence of a nucleation site, while the MS IBI potential remains fluid (Fig. 6). Note, for a direct comparison with the MS IBI model derived here, antifreeze particles were not used with the MARTINI model. To ensure that the MS IBI system is not an amorphous solid structure, the ratio of the diffusion coefficients with and without a nucleation site were calculated for each model from the slope of the mean-squared displacement. As shown in Table 2, the diffusion coefficient of the MS IBI potential model remains relatively unchanged when a nucleation site is added, whereas a significant drop is seen for the MARTINI model resulting from crystallization. Additionally, Fig. 7 plots the RDF of the MARTINI model for the bulk NVT state as compared to the 4:1 mapped target data. Clearly, the MARTINI potential does not accurately capture the

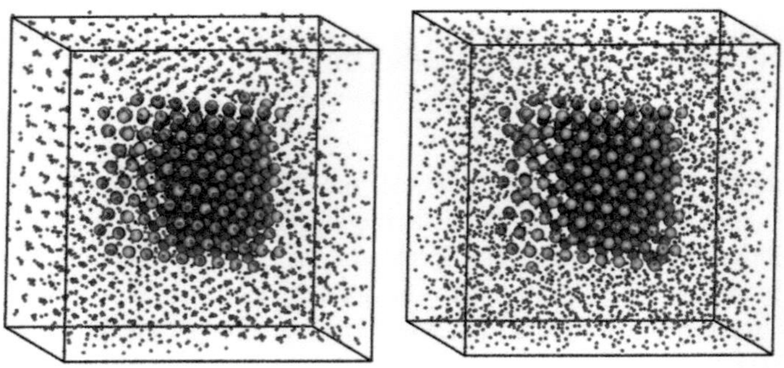

Fig. 6 Configurations from simulations in the presence of a nucleation site with the MARTINI (*left*) and MS IBI (*right*) models. CG water beads colored *silver* were kept fixed during the simulations, but were treated as the same type as the *blue* particles (i.e., the color is different to show the nucleation site)

Table 2 Ratio of diffusion coefficients from simulations with (D_{nuc}) and without (D_{bulk}) a nucleation site with different potentials

Model	D_{nuc}/D_{bulk}
MS IBI	0.88
MARTINI	0.02

Diffusion coefficient D calculated from the slope of a linear fit to the long-time mean squared displacement (MSD), using MSD = 6Dt

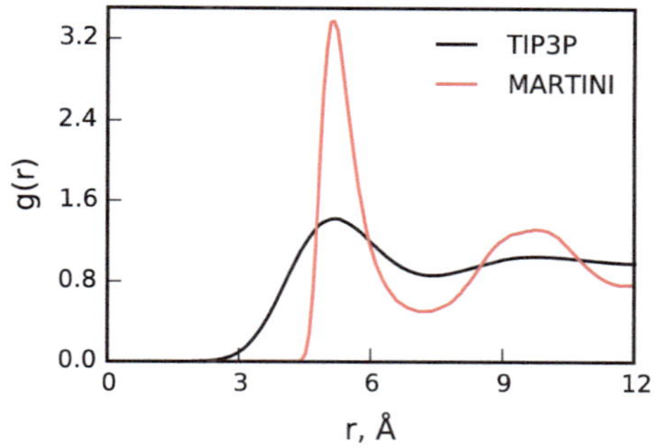

Fig. 7 RDFs of the MARTINI model and the atomistic TIP3P model mapped to the CG level for the bulk NVT state

structural correlations of bulk water, further demonstrating the significant improvement of the MS IBI model in reproducing key properties of water.

We note that the self-diffusion coefficient of MS IBI water is calculated to be 16.07×10^{-9} m^2/s at 305 K and 1 atm, as compared to 3.05×10^{-9} m^2/s for the atomistic TIP3P water at the same conditions, both run for 5 ns. This factor of ~ 5 difference is not entirely unexpected, given the softening of the free energy landscape that often comes with CG models and the fact that kinetic data was not used in the optimization. However, we also note that the dynamics of the CG model does not bear a strong connection with the atomistic level behavior, given that each CG bead represents 4 water molecules, but not necessarily the same water molecules through time, due to the lack of permanent bonds between the waters being grouped together.

4 Conclusions

In this work, the MS IBI method was used to derive the interactions for a 4:1 mapped CG water model, using a modified version of the Chiu, et al. potential as an initial guess. An improvement over previous models is made by simultaneously matching the fluid structure to target data from bulk and interfacial states. It was shown that a model that reproduces the structure and density of water does not necessarily reproduce the interfacial properties and that the addition of a droplet target state constrains the potential to also capture the interfacial properties. The resulting potential is able to accurately predict the density of water at 305 K and 1 atm, interfacial properties, and structural correlations. Additionally, the model shows no tendency to spontaneously crystallize at physiological conditions. This is important, since inaccuracies in a water model can propagate as more potentials are derived against it when simulating mixed systems.

This work highlights a key advantage of deriving potentials via the MS IBI approach. For simulations that cover multiple states, it is important to have a forcefield that is accurate across the states of interest. MS IBI allows this to be achieved by including target data from states that represent structures present in the states of interest. This is realized here by including a multi-ensemble state to accurately model the pressure-density relationship, and a droplet state to capture the interfacial properties of water. Another case where this would be beneficial is studying systems over multiple phases, e.g., phase transitions in liquid crystals. While clever approaches are taken to capture behavior across multiple states [43], a more systematic approach would be useful. Based on the results presented here, we foresee this method being useful for deriving CG potentials for a wide range of applications.

Acknowledgments Funding was provided by Grant No. R01AR057886-01 from the National Institute of Arthritis and Musculoskeletal and Skin Diseases and the National Science Foundation under Grant OCI-0904879.

References

1. Wang, Y., Voth, G.A.: Unique spatial heterogeneity in ionic liquids. J. Am. Chem. Soc. **127**, 12192–12193 (2005)
2. Bhargava, B.L., DeVane, R., Klein, M.L., Balasubramanian, S.: Nanoscale organization in room temperature ionic liquids: a coarse grained molecular dynamics simulation study. Soft Matter **3**, 1395–1400 (2007)
3. Karimi-Varzaneh, H.A., Müller-Plathe, F., Balasubramanian, S., Carbone, P.: Studying long-time dynamics of imidazolium-based ionic liquids with a systematically coarse-grained model. Phys. Chem. Chem. Phys. **12**, 4714–4724 (2010)
4. Padding, J., Briels, W.: Time and length scales of polymer melts studied by coarse-grained molecular dynamics simulations. J. Chem. Phys. **117**, 925–943 (2002)
5. Harmandaris, V.A., Floudas, G., Kremer, K.: Temperature and pressure dependence of polystyrene dynamics through molecular dynamics simulations and experiments. Macromolecules **44**, 393–402 (2010)
6. Sun, Q., Faller, R.: Crossover from unentangled to entangled dynamics in a systematically coarse-grained polystyrene melt. Macromolecules **39**, 812–820 (2006)
7. Milano, G., Müller-Plathe, F.: Mapping atomistic simulations to mesoscopic models: a systematic coarse-graining procedure for vinyl polymer chains. J. Phys. Chem. B **109**, 18609–18619 (2005)
8. Shinoda, W., DeVane, R., Klein, M.L.: Coarse-grained molecular modeling of non-ionic surfactant self-assembly. Soft Matter **4**, 2454–2462 (2008)
9. Lee, H., Pastor, R.W.: Coarse-grained model for PEGylated lipids: effect of PEGylation on the size and shape of self-assembled structures. J. Phys. Chem. B **115**, 7830–7837 (2011)
10. Srinivas, G., Discher, D.E., Klein, M.L.: Self-assembly and properties of diblock copolymers by coarse-grain molecular dynamics. Nat. Mater. **3**, 638–644 (2004)
11. Nguyen, H.D., Hall, C.K.: Molecular dynamics simulations of spontaneous fibril formation by random-coil peptides. Proc. Natl. Acad. Sci. U.S.A. **101**, 16180–16185 (2004)
12. Iacovella, C.R., Keys, A.S., Glotzer, S.C.: Self-assembly of soft-matter quasicrystals and their approximants. Proc. Natl. Acad. Sci. U.S.A. **108**, 20935–20940 (2011)
13. Hadley, K.R., McCabe, C.: Coarse-grained molecular models of water: a review. Mol. Simul. **38**, 671–681 (2012)
14. Basdevant, N., Borgis, D., Ha-Duong, T.: A semi-implicit solvent model for the simulation of peptides and proteins. J. Comput. Chem. **25**, 1015–1029 (2004)

15. Basdevant, N., Ha-Duong, T., Borgis, D.: Particle-based implicit solvent model for biosimulations: application to proteins and nucleic acids hydration. J. Chem. Theory Comput. **2**, 1646–1656 (2006)
16. Masella, M., Borgis, D., Cuniasse, P.: Combining a polarizable force-field and a coarse-grained polarizable solvent model. II. accounting for hydrophobic effects. J. Comput. Chem. **32**, 2664–2678 (2011)
17. Marrink, S.J., Risselada, H.J., Yefimov, S., Tieleman, D.P., de Vries, A.H.: The MARTINI force field: coarse grained model for biomolecular simulations. J. Phys. Chem. B **111**, 7812–7824 (2007)
18. Chiu, S.-W., Scott, H.L., Jakobsson, E.: A coarse-grained model based on morse potential for water and n-alkanes. J. Chem. Theory Comput. **6**, 851–863 (2010)
19. Shinoda, W., DeVane, R., Klein, M.L.: Multi-property fitting and parameterization of a coarse grained model for aqueous surfactants. Mol. Simul. **33**, 27–36 (2007)
20. Hadley, K.R., McCabe, C.: On the investigation of coarse-grained models for water: balancing computational efficiency and the retention of structural properties. J. Phys. Chem. B **114**, 4590–4599 (2010)
21. Van Hoof, B., Markvoort, A.J., Van Santen, R.a.; Hilbers, P.a.J.: The CUMULUS coarse graining method: transferable potentials for water and solutes. J. Phys. Chem. B **115**, 10001–10012 (2011)
22. Reith, D., Pütz, M., Müller-Plathe, F.: Deriving effective mesoscale potentials from atomistic simulations. J. Comput. Chem. **24**, 1624–1636 (2003)
23. Moore, T.C., Iacovella, C.R., McCabe, C.: Derivation of coarse-grained potentials via multistate iterative Boltzmann inversion. J. Chem. Phys. **140**, 224104 (2014)
24. Hadley, K.R., McCabe, C.: A coarse-grained model for amorphous and crystalline fatty acids. J. Chem. Phys. **132**, 134–505 (2010)
25. Bayramoglu, B., Faller, R.: Coarse-grained modeling of polystyrene in various environments by iterative Boltzmann inversion. Macromolecules **45**, 9205–9219 (2012)
26. Qian, H.J., et al.: Temperature-transferable coarse-grained potentials for ethylbenzene, polystyrene, and their mixtures. Macromolecules **41**, 9919–9929 (2008)
27. Bayramoglu, B., Faller, R.: Modeling of polystyrene under confinement: exploring the limits of iterative boltzmann inversion. Macromolecules **46**, 7957–7976 (2013)
28. Carbone, P., et al.: Transferability of coarse-grained force fields: the polymer case. J. Chem. Phys. **128**, 064904 (2008)
29. Hartigan, J.A., Wong, M.A.: Algorithm AS 136: a k-means clustering algorithm. Appl. Stat. **28**, 100–108 (1979)
30. Jorgensen, W.L., Chandrasekhar, J., Madura, J.D., Impey, R.W., Klein, M.L.: Comparison of simple potential functions for simulating liquid water. J. Chem. Phys. **79**, 926–935 (1983)
31. Plimpton, S.: Fast parallel algorithms for short-range molecular dynamics. J. Comput. Phys. **117**, 1–19 (1995)
32. LAMMPS WWW Site—http://lammps.sandia.gov, http://lammps.sandia.gov
33. A git repository for this package is hosted at https://github.com/ctk3b/msibi, http://github.com/ctk3b/msibi
34. Anderson, J.A., Lorenz, C.D., Travesset, A.: General purpose molecular dynamics simulations fully implemented on graphics processing units. J. Comput. Phys. **227**, 5342–5359 (2008)
35. Glaser, J., Nguyen, T.D., Anderson, J.A., Lui, P., Spiga, F., Millan, J.A., Morse, D.C., Glotzer, S.C.: Strong scaling of general-purpose molecular dynamics simulations on GPUs. Comput. Phys. Commun. **192**, 97–107 (2015)
36. HOOMD-Blue web page. http://codeblue.umich.edu/hoomd-blue, http://codeblue.umich.edu/hoomd-blue
37. McGibbon, R.T., Beauchamp, K.A., Schwantes, C.R., Wang, L.-P., Hernández, C.X., Harrigan, M.P., Lane, T.J., Swails, J.M., Pande, V.S.: MDTraj: a modern, open library for the analysis of molecular dynamics trajectories. bioRxiv 2014
38. A Git repository for this package is hosted at https://github.com/mdtraj/mdtraj. http://github.com/ctk3b/msibi

39. Vega, C., de Miguel, E.: Surface tension of the most popular models of water by using the test-area simulation method. J. Chem. Phys. **126**, 154707 (2007)
40. Wang, H., Junghans, C., Kremer, K.: Comparative atomistic and coarse-grained study of water: what do we lose by coarse-graining? Eur. Phys. J. E **28**, 221–229 (2009)
41. Yesylevskyy, S.O., Schafer, L.V., Sengupta, D., Marrink, S.J.: Polarizable water model for the coarse-grained MARTINI force field. PLoS Comput. Biol. **6**, e1000810 (2010)
42. Zavadlav, J., Melo, M.N., Marrink, S.J., Praprotnik, M.: Adaptive resolution simulation of polarizable supramolecular coarse-grained water models. J. Chem. Phys. **142**, 244118 (2015)
43. Mukherjee, B., Delle Site, L., Kremer, K., Peter, C.: Derivation of coarse grained models for multiscale simulation of liquid crystalline phase transitions. J. Phys. Chem. B **116**, 8474–8484 (2012)

Optimizing Molecular Models Through Force-Field Parameterization via the Efficient Combination of Modular Program Packages

Marco Hülsmann, Karl N. Kirschner, Andreas Krämer, Doron D. Heinrich, Ottmar Krämer-Fuhrmann and Dirk Reith

Abstract A central goal of molecular simulations is to predict physical or chemical properties such that costly and elaborate experiments can be minimized. The reliable generation of molecular models is a critical issue to do so. Hence, striving for semiautomated and fully automated parameterization of entire force fields for molecular simulations, the authors developed several modular program packages in recent years. The programs run with limited user interactions and can be executed in parallel on modern computer clusters. Various interlinked resolutions of molecular modeling are addressed: For intramolecular interactions, a force-field optimization package named $Wolf_2Pack$ has been developed that transfers knowledge gained from quantum mechanics to Newtonian-based molecular models. For intermolecular interactions, especially Lennard–Jones parameters, a modular optimization toolkit of programs and scripts has been created combining global and local optimization algorithms. Global optimization is performed by a tool named CoSMoS, while local optimization is done by the gradient-based optimization workflow named GROW or by a derivative-free method called SpaGrOW. The overall goal of all program packages is to realize an easy, efficient, and user-friendly development of reliable force-field parameters in a reasonable time. The various tools are needed

M. Hülsmann (✉) · A. Krämer · D.D. Heinrich · D. Reith
Department of Mechanical Engineering (EMT) and Institute for Technology, Renewables and Energy Efficiency (TREE), Bonn-Rhein-Sieg University of Applied Sciences, Grantham-Allee 20, 53757 Sankt Augustin, Germany
e-mail: marco.huelsmann@h-brs.de; marco.huelsmann@scai.fraunhofer.de

M. Hülsmann · O. Krämer-Fuhrmann · D. Reith
Department of Simulation Engineering, Fraunhofer-Institute for Algorithms and Scientific Computing (SCAI), Schloss Birlinghoven, 53757 Sankt Augustin, Germany

K.N. Kirschner
Department of Computer Science and the Institute of Visual Computing (IVC), Bonn-Rhein-Sieg University of Applied Sciences, Grantham-Allee 20, 53757 Sankt Augustin, Germany

R.Q. Snurr et al. (eds.), *Foundations of Molecular Modeling and Simulation*, Molecular Modeling and Simulation, DOI 10.1007/978-981-10-1128-3_4

and interlinked since different stages of the optimization process demand different courses of action. In this paper, the conception of all programs involved is presented and how they communicate with each other.

Keywords Molecular modeling · Force field · Numerical optimization · High-performance computing · Modular software packages

1 Introduction

1.1 Molecular Simulation and Its Tools

Molecular simulation methods, most prominently molecular dynamics (MD) and Monte Carlo (MC), are powerful tools to gain insight into microscopic processes that govern the macroscopic behavior of matter. There is a long-standing tradition of studying molecular behavior for biomolecules (e.g., proteins, DNA, and carbohydrates) and for soft materials (e.g., plastics, fibers, carbon nanotubes, and ionic liquids). This is reflected by a long history of parameter and software development in this area, which is often distributed together as a collection of predefined parameters, molecular building blocks, and a simulation engine. However, in recent years, significant algorithmic progress has been made to enhance molecular simulation and analysis. There is a widespread utilization of GPUs in existing software packages (e.g., *Amber* [1], *Charmm* [2], *Gromacs* [3], and *LAMMPS* [4]) and automated procedures to derive force-field parameters [5, 6]. In addition, recent coarse-grained methods that access the mesoscale introduced new powerful scientific concepts to the field of molecular simulations (e.g., *HOOMD* [7], *ESPResSo*++ [8], and *IBIsCO* [9]).

To gain a molecular-level understanding, chemical systems are modeled at atomistic or near atomistic (e.g., united atom, fine coarse graining) resolution levels. Since computable properties obey the laws of statistical physics, an ensemble of several ten thousands of atoms is necessary to compute the macroscopic observables. Furthermore, modern industrially relevant systems (e.g., chemically heterogeneous, surfaces, mixed phase states) require large models for accurate representations. This results in the necessity to implement the calculations in high-performance computing environments. Driven by the ongoing growth in computational power, it can be expected that these molecular methods will be increasingly useful in the coming decades.

One goal of our research is to provide a computational modeling service to external researchers, both in industry and academics, who wish to obtain a molecular understanding of their systems. As such, we have been faced with using, modifying, and optimizing all atom, united atom, and coarse-grained force fields for natural products, polymers, lipids, ionic liquids, and organic solvents. While the technique of molecular simulations has existed for decades and in spite of its obvious powers, only a few companies have in-house departments, that is due to

(a) the diversity of knowledge needed to do high-quality research (i.e., the method's core is mathematics and physics, the content is often being chemical, and the technical aspects require computational scientists) and (b) the high-performance hardware that is required to execute the simulation software.

1.2 Force Fields

One key requirement in molecular mechanics (MM)-based models is the need to be as accurate as possible. This accuracy is directly dependent upon the force field, which describes the intra- and intermolecular interactions. Force fields are a semiempirical approach to represent these interactions—that is a set of equations and associated parameters that model stretching, bending, internal rotations, van der Waals, and Coulombic interactions. In general, there is a consensus on what function form of the equations should be used. Coupled directly to the equations are the parameters, whose optimization is very important but often tedious to accomplish.

Over the past decades, many researchers have developed force fields for a variety of areas, such as thermodynamic properties of fluids [10–15], mechanic properties of solids [16–18], phase change phenomena [19–21], protein folding [22–24], transport processes in biological tissue [25, 26], transport processes in liquids [27–29], polymer properties using different length scales [30–33], and generic statistic properties of soft matter [34]. Some of these force fields have been molecule specific, while others have been transferable over a chemical class (e.g., hydrocarbons, alcohols). For our models, the criterion is that they accurately reproduce or predict the relevant observable(s) using the modeling software that is most appropriate for the investigation. Quantum mechanical methods are useful to determine some of the target observables used in parameter fitting (i.e., geometry, electrostatics, relative energies). However, weak short-range nonbonded interactions are difficult to isolate target quantum mechanical observables, particularly when the molecules are composed of heterogeneous atom types. Hence, the force-field parameters for these weak interactions are often fitted to experimental condense-phase target values. Thus, a manual parameter adjustment is usually not feasible or is, at best, extremely time-consuming.

1.3 Goal of This Work

What has become clear is that a user-friendly and versatile software package, which facilitates the optimization of force-field parameters for a given MM or MD engine, is very important. Hence, automated and semiautomated parameterization process can reduce the time required for optimization and subsequently allow researchers more time to explore their ideas. We contribute to this field by creating modular software packages that follow our ideas for force-field development and by efficiently and systematically combining these programs for the (semi)automated optimization of bonded and nonbonded parameters.

The benefits of utilizing scientific workflows are numerous, and they represent a major improvement in how one approaches force-field development. These benefits include (a) saving time by automating certain optimization tasks; (b) making force-field development quasi-deterministic; (c) reducing human error; (d) enabling tasks to be executed in a distributed environment; (e) accommodating ideas, algorithmic changes, and updates easier; and finally (f) accelerating and transforming the process of scientific analysis. From a scientific perspective, workflows enable researchers to focus more on scientific issues, and due to its hierarchical organization, new advancement in theories can be easily incorporated. In addition to this, errors within the force field and models are better avoided, making the simulation results become more trustworthy and reliable. Moreover, the algorithms involved within the workflow can handle overdetermined and underdetermined optimization problems. From a community service perspective, our workflows significantly reduce the real time needed for force-field development and allow nonspecialists access to more standardized optimization procedures.

For the determination of the intramolecular parameters, we developed a tool named $\mathrm{Wolf_2Pack}$, and for the intermolecular parameters, we use a combination of a global optimization procedure with a local one. For the former, we developed a global optimization tool named CoSMoS, and for the latter, we developed a gradient-based optimization toolkit named GROW and a derivative-free sparse grid-based algorithm named SpaGrOW. The three tools are described in more detail in the next subsections.

2 Goal-Driven Software Conception

2.1 *Wolf$_2$Pack: Intramolecular Parameters*

The concept for $\mathrm{Wolf_2Pack}$[1] came from our goals to have a tool that would

(a) allow for quick optimization of bonded parameters,
(b) enable one to qualify observed MD structural results,
(c) allow one to evaluate existing force fields,
(d) allow for the systematic generation and archiving of QM target data for reuse,
(e) enable nonforce-field experts the opportunity to generate their own parameters, and
(f) enable reproducibility of reported force-field research results (e.g., molecule-specific QM and MM energy curves).

[1]http://www.wolf2pack.com.

To achieve these goals, a scientific workflow was developed that provided a guiding architecture for software development [35]. Each step of the workflow was realized through shell scripts, whose output data are organized, as illustrated in Fig. 1, into subdirectories. This modular construct has the advantage that individual scripts can be easily updated, discovered errors in the scripts and generated data can be efficiently corrected, and the generated data are organized in a systematic manner that easily allow for the inclusion of new computations, archiving, and reuse.

To enable nonforce-field experts the chance to check and optimize parameters, a Web site was created that serves as a front-end to Wolf$_2$Pack [36]. This Web site guides users in the parameter optimization process, starting from selecting an appropriate molecule to the determination of a suitable parameter. The site also provides a collection of "Knowledge Modules" that are a combination of tutorials and examples. Currently, the Web site only provides access to a truncated amount of the existing data within the Wolf$_2$Pack's database. In the near future, we intend to provide users' access to the full database and enable them to upload a molecule and compute the QM curves that they desire.

An important component of Wolf$_2$Pack is its molecular database. The database contains molecules of diverse chemical functionalities for which bond, angle, and torsion relative energies curves have been generated. This database naturally grows over time as new functional groups and combinations thereof are investigated. Thus, the statistical evaluation of force fields improves as the database expands. Due to its systematic development, the database also enables users to reproduce results in published force-field papers, which is currently a difficult task to accomplish. We believe this will become an important feature in the future as users make use of Wolf$_2$Pack for optimizing parameters. The challenge will be to continually update the database for the new QM theories that are reported in the

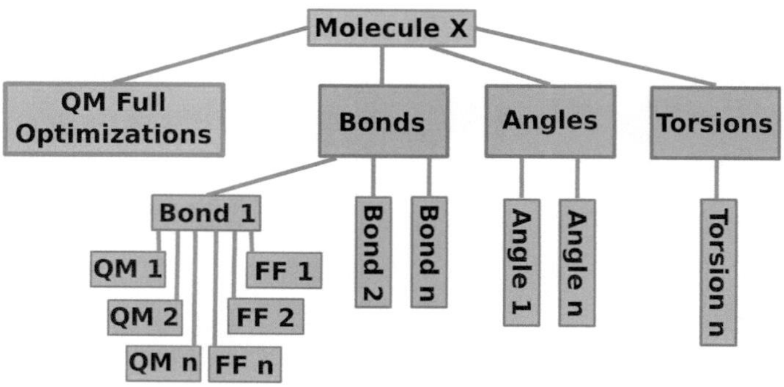

Fig. 1 Illustration of the basic directory structure within Wolf$_2$Pack. Each molecule with a given conformation has its own parent directory. The *number* of bond, angle, and torsion subdirectories is dependent upon the molecule's unique internal coordinates. The "QM n" and "FF n" labels indicate data from constraint QM and MM optimizations using a specific theory level (e.g., HF/6-31G(d)//HF/6-31G(d)) or force field (e.g., Parm14SB)

literature, which will be an increasingly demanding task as the number of molecules and internal coordinates grow.

Considering parameterization philosophy, we are pursuing new ideas in addition to the traditional fitting of continuous relative potential energy curves. Through the assistance of the Balloon algorithm [37], Wolf$_2$Pack can quantum mechanically generate and identify unique conformations automatically. For illustration, we recently predicted 76 unique octane conformations at the HF/6-31G(d) using Balloon and Wolf$_2$Pack algorithms. While this does not represent the complete set of unique octane conformations, which have been determined to be 95 [38], it does impressively cover a wide range of relative energies (0.0–8.9 kcal/mol). These high numbers of conformations for a flexible molecule allow for a unique way to validate force fields. Traditionally, nonbonded and bonded force-field terms are optimized by reproducing experimental observables (e.g., density) and relative energy curves (i.e., transition states, minima), which rarely consider more than a few high energy minima. By having access to a large number of minima, one can observe how a given force field's parameters transfer to higher energy minima and conformations not originally considered during the optimization process.

Researchers usually strive to generate continuous QM rotational energy curves. A continuous curve is one whose incremented internal coordinate changes, while all other unconstrained torsion angles remain in their original position (e.g., within ±5°). The advantage of this is that the obtained relative energies directly reflect the rotation around a single bond. The subsequent parameter optimization is then fairly straightforward. A discontinuous rotational curve would be when a second torsion undergoes significant rotation at some point during the interested torsion rotation (e.g., Fig. 2). The resulting energy curve then reflects contribution from changes

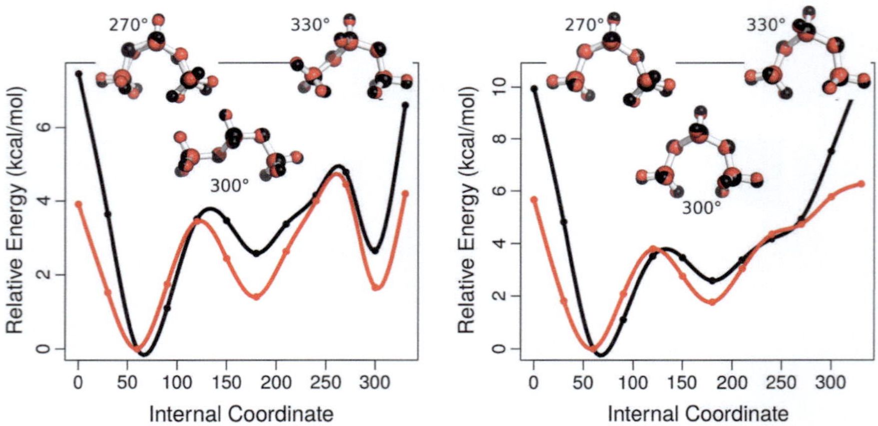

Fig. 2 Potential energy curves and geometric overlays for dimethoxymethane as determined by HF/6-31G(d) (*red*) and the Gaff (*black*) force field. In this case, the C–C–O–C torsion on the *left side* of the molecule is systematically rotated. The *left image* shows the discontinuous curve where the *right side* C–O–C–C adopted a transconformation at 300°, while the *right image* shows the continuous curve. The continuous curve was generated by constraining the mobile torsion

within two torsion angles, making parameter optimization more convoluted. In $Wolf_2Pack$, we strive to generate continuous curves and will apply a secondary torsion constraint if necessary to obtain one for parameter optimization purposes. Nevertheless, we also make use of the discontinuous curves that are produced for testing the robustness of the optimized parameters. Fundamentally, the discontinuous curve represents significant coupling between internal coordinates, for which force fields should ideally reproduce. We believe that reproduction of discontinuous curves is a more rigorous test of a force field's performance in comparison with the reproduction continuous curves. In addition to investigated torsion angles, discontinuous curves also occur when generating bond stretching and angle bending energy profiles. Typically, a close contact occurs between atoms, resulting in the rotation about a bond to relieve the high energy strain.

2.2 *CoSMoS, GROW, and SpaGrOW: Intermolecular Parameters*

The optimization of nonbonded parameters is difficult since one can rarely isolate the parameters for a specific atom type, with the notable exception of the noble gases. If one considers simple saturated hydrocarbons, the carbon and hydrogen Lennard–Jones parameters are often optimized simultaneously. This results in a large possible parameter space, making an a priori understanding of the loss function's shape impossible. For this reason, as illustrated in Fig. 3, we have developed both global (i.e.,

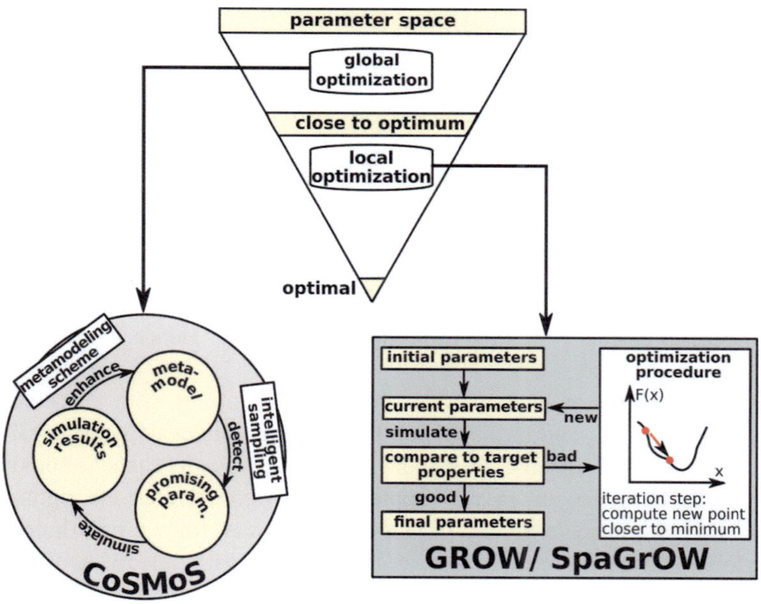

Fig. 3 The funnel workflow approach for optimizing nonbonded parameters

CoSMoS) and local (i.e., GROW and SpaGrOW) tools that are implemented in a funnel workflow. CoSMoS is based on metamodeling that enables rough identification of potential optimal values, while either a gradient-based (GROW) or derivative-free (SpaGrOW) approach is used to refine the identified parameters.

In the last two decades, substantial research occurred for the optimization of intermolecular force-field parameters [39–54]. In most cases, intermolecular parameters, especially Lennard–Jones parameters, cannot be strictly derived via physical considerations since they parameterize semiempirical models (i.e., based on classical mechanics) whom themselves only approximate reality. Hence, they are usually adjusted so that the resulting model is able to reproduce physical or chemical experimental target properties as accurately as possible.

The overall optimization task is to find a solution to the following mathematical optimization problem:

$$\min_{x\in\Omega} F(x) := \left\| W(f^{\text{sim}}(x) - f^{\text{exp}}) \right\|_p^2, \; p \in [1, \infty], \tag{1}$$

where $x = (x_1, \ldots, x_N)^T \in \mathbb{R}^N$ is a vector consisting of the force-field parameters to be adjusted, $N \in \mathbb{N}$ is the number of parameters, $n \in \mathbb{N}$ is the number of physical properties to be fitted, $f^{\text{sim}}(x) \in \mathbb{R}^n$ is the vector containing all properties calculated by simulation, f_i^{sim}, $i = 1, \ldots, m$, and $f^{\text{exp}} \in \mathbb{R}^n$ is the vector containing the experimental target values f_i^{exp}, $i = 1, \ldots, m$. For reasons of brevity, $\|\cdot\|$ indicates an arbitrary $p \in [1, \infty]$. If a particular norm is considered, this will be expressed explicitly (e.g., $\|\cdot\|_2$ or $\|\cdot\|_\infty$). The weighting matrix is defined as:

$$W = \begin{pmatrix} \frac{w_1}{f_1^{\text{exp}}} & 0 & \cdots & 0 \\ 0 & \frac{w_2}{f_2^{\text{exp}}} & \ddots & \vdots \\ \vdots & \ddots & \ddots & 0 \\ 0 & \cdots & 0 & \frac{w_n}{f_n^{\text{exp}}} \end{pmatrix} \tag{2}$$

with specific weights w_i, $i = 1, \ldots, n$, for each property, accounting for the fact that some properties may be easier to reproduce than others due to statistical noise on both simulation and experimental data. The loss function $F(x)$ has to be minimized with respect to x within an admissible domain $\Omega \subset \mathbb{R}^N$. Hence, the optimization problem is constrained.

The loss function does not have any analytical form with respect to the force-field parameters, and the simulated properties are affected by statistical noise. Hence, it cannot be assumed to be smooth or differentiable. Its shape is not known a priori and is often jagged in real applications. Moreover, as the optimization problem may be overdetermined, the loss function may form a rain drain, where many global optima are located at the bottom. Additionally, the evaluations of the loss function may be costly, in particular if molecular simulations have to be performed. For all these reasons, the solution of the optimization problem (1) is challenging and not possible using standard line-search methods. In order to jump over intermediate local minima, an efficient global optimization that focuses into a close neighborhood of the global minimum is indispensable. Mostly, global optimization algorithms get stuck at a certain iteration because the points in the parameter space are generated via random sampling methods. In this case, local optimization procedures are more reliable and faster because they are directed to the minimum, especially when they are gradient based. Hence, the combination of global with local optimization algorithms turned out to be much more reliable and efficient in order to solve the present optimization task than the usage of a single global or local algorithm [55].

2.3 *Methodological Aspects of CoSMoS*

The recently developed global optimization tool for the Calibration of molecular force fields by Simultaneous Modeling of Simulated data (CoSMoS) [56] uses a metamodeling procedure based on radial basis functions (RBFs). It has been shown in [56] that metamodel-based optimizers particularly suit the quest for quickly finding nearly optimal force-field parameters. The metamodels constructed by CoSMoS describe functional dependencies between the force-field parameters and the relative deviations of the simulated properties to experimental data so that the minimization task is easier to solve. The RBFs are rational symmetric functions $\Phi : \mathbb{R}^N \to \mathbb{R}$ of the form $\Phi(x) = \Phi(\|x\|)$ for $x \in \mathbb{R}^N$. For the present optimization problem, inverse multiquadric RBFs, i.e., $\Phi(x) = (\|x\|^2 + \gamma^2)^{-\frac{1}{2}}$, $\gamma \in \mathbb{R}$, turned out to perform best. However, CoSMoS also offers the possibility to use other RBFs, e.g., cubic $\left(\Phi(x) = \|x\|^3\right)$ and Gaussian $(\Phi(x) = \exp(-(\gamma\|x\|)^2))$ functions, thin-plate splines $(\Phi(x) = \|x\|^2 \log\|x\|)$, or multiquadrics $\left(\Phi(x) = \sqrt{\|x\|^2 + \gamma^2}\right)$. The metamodel $\mathcal{M}^v(x)$ interpolating a target property $v \in \{1, \ldots, n\}$ is then given by

$$\mathcal{M}^v(x) = \sum_{j=1}^{q} \alpha_j^v \Phi(\|x - x_j\|) + \sum_{k=1}^{r} \beta_k^v p_k(x), \tag{3}$$

where x_j, $j = 1, \ldots, q$, $q \in \mathbb{N}$ are sampling points that fulfill the interpolation condition $\mathcal{M}^v(x_j) = f_v^{\text{sim}}(x_j)$, $j = 1, \ldots, q$. The $p_k(x)$, $k = 1, \ldots, r$, $r \in \mathbb{N}$ are low-order polynomials, and the coefficients $\alpha_j^v \in \mathbb{R}$, $j = 1, \ldots, q$, $v = 1, \ldots, n$ and $\beta_k^v \in \mathbb{R}$, $k = 1, \ldots, r$, $v = 1, \ldots, n$ are obtained by solving a linear equation system (LES): The radial basis function matrix of the sampling points is given by $H = (H)_{li} := (\Phi(\|x_l - x_i\|))_{l,i=1,\ldots,q} \in \mathbb{R}^{q \times q}$, and the polynomial matrix is given by $P := (P)_{lk} = p_k(x_l)_{l=1,\ldots,q,k=1,\ldots,r} \in \mathbb{R}^{q \times r}$. The right hand side is as follows:

$$d_v^{\text{sim}} := (d_v^{\text{sim}})_l = \left(\frac{f_v^{\text{sim}}(x_l) - f_v^{\text{exp}}}{s_v^{\text{sim}} f_v^{\text{exp}}} \right)_{l=1,\ldots,q}, \tag{4}$$

where $s_v^{\text{sim}}, v \in \{1, \ldots, n\}$ is the standard deviation of the relative noise of the property v. Hence, the following linear equation system (LES) has to be solved:

$$\begin{pmatrix} \text{H} & \text{P} \\ P^T & 0 \end{pmatrix} \begin{pmatrix} \alpha^v \\ \beta^v \end{pmatrix} = \begin{pmatrix} d^{\text{sim}} \\ 0 \end{pmatrix}, \tag{5}$$

where $\begin{pmatrix} \alpha^v \\ \beta^v \end{pmatrix}$ is the vector containing the coefficients $\alpha_j^v \in \mathbb{R}$, $j = 1, \ldots, q$, $v = 1, \ldots, n$, and $\beta_k^v \in \mathbb{R}$, $k = 1, \ldots, r$, $v = 1, \ldots, n$. The second line mirrors an additional orthogonality to render the coefficients unique. However, this procedure may lead to large RBF coefficients, resulting in wavy metamodels that do not reflect the underlying data properly. This is particularly severe for noisy data, which demands proper smoothing approaches. Thus, in this work, CoSMoS was extended by two different smoothing methods: The *smoothest* metamodel is the one with the smallest RBF coefficients, which can be calculated by solving

$$\min_{\alpha^v} \quad \|\alpha^v\|^2, \tag{6}$$

$$\text{where} \quad f_l^{\text{sim}} - \xi \leq b_l \leq f_l^{\text{sim}} + \xi, l = 1, \ldots, q, \tag{7}$$

where $\xi > 0$ is a small tolerance value, and b is the vector $(\text{H} \quad \text{P}) \begin{pmatrix} \alpha^v \\ \beta^v \end{pmatrix}$. As the statistical noise is taken into account by the method due to Eq. (4), confidence intervals are drawn around the sampling points so that overfitting can be avoided during interpolation. Hence, the method searches for metamodels which are as smooth as possible.

The *weighted* smoothing method tries to find a compromise between the two contradictory requirements of high smoothness and low smoothing error. This compromise is controlled via an additional weighting parameter $\chi > 0$, and the following constrained minimization problem is solved:

$$\min_{\alpha^v, \beta^v} \left\| (\text{H} \quad \text{P}) \begin{pmatrix} \alpha^v \\ \beta^v \end{pmatrix} - d_v^{\text{sim}} \right\|^2 + \chi \|\alpha^v\|^2, \tag{8}$$

which is equivalent to solving the LES

$$\begin{pmatrix} H^T H + \chi I & H^T P \\ P^T H & P^T P \end{pmatrix} \begin{pmatrix} \alpha^v \\ \beta^v \end{pmatrix} = \begin{pmatrix} H^T d_v^{\text{sim}} \\ P^T d_v^{\text{sim}} \end{pmatrix}. \tag{9}$$

An optimal choice of χ would lead to a perfect metamodel fulfilling both criteria. However, the parameter is problem-dependent and thus difficult to optimize in practice.

Furthermore, CoSMoS provides an intelligent sampling procedure extending the approach of the Constrained Optimization using Response Surfaces (CORS) [57]. The latter focuses the sampling onto potentially optimal regions, avoiding previously sampled regions. This neighborhood is a ball around a sampling point $x \in \tilde{\Omega}$, where $\tilde{\Omega} \subset \Omega$ is the set of the already sampled points, of radius

$$r < \delta_{\tilde{\Omega}}^{\max} := \max_{x \in \Omega} \min_{\tilde{x} \in \tilde{\Omega}} \|x - \tilde{x}\|. \tag{10}$$

This taboo search approach is then realized by solving the constrained minimization problems:

$$\min_{x \in \Omega} \quad ||W \cdot \mathcal{M}^{v}(x)||, \tag{11}$$

$$\text{where} \quad x \in \bigcup_{\tilde{x} \in \tilde{\Omega}} U_r(\tilde{x}), \quad v = 1, \ldots, n. \tag{12}$$

CoSMoS extends this approach by introducing a penalty term

$$p(x) := \frac{\delta_{\tilde{\Omega}}^{\max}}{\min_{\tilde{x} \in \tilde{\Omega}} \|x - \tilde{x}\|} \geq 1, \tag{13}$$

which grows to infinity, whenever x approaches a sampling point. In contrast to CORS, CoSMoS minimizes the penalized metamodels

$$\tau_{\tilde{\gamma}}^{v}(x) := p(x)^{\tilde{\gamma}} (\mathcal{M}^{v}(x) - c), v = 1, \ldots, n. \tag{14}$$

where $\tilde{\gamma}$ and c are control parameters. For more algorithmic details, see reference [56]. Figure 4 demonstrates the adaptive nature of the intelligent sampling strategy. The plot shows a preliminary metamodel after 20 evaluations (right) compared to the actual loss function (left). The metamodel generally captures the optimal region of the loss function, i.e., the vicinity of the minimum. The intelligent sampling strategy takes advantage of this and preferably samples points in the optimal region. In return, each function evaluation further improves the accuracy of the metamodel, improving the sketch of the optimal region. This circular procedure within CoSMoS, which is also depicted in Fig. 3, reduces the number of required simulations and thus the time-to-solution substantially.

An additional advantage of CoSMoS is the fact that it can handle abortive simulations. Whenever a simulation goes wrong due to a bad selection of the force-field parameters, the corresponding sampling points are penalized in the same way so that they are not triggered anymore by the sampling algorithm. Within one

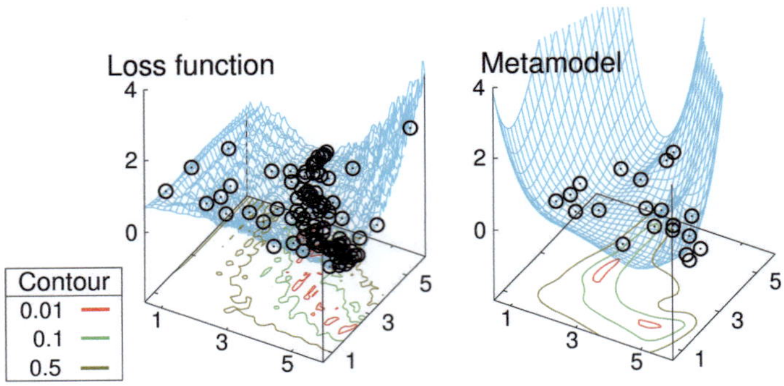

Fig. 4 *Left* The original loss function for a test problem is shown. The *black points*, sampled by CoSMoS, adapt the shape of the loss function. *Right* The metamodel of the loss function after 20 CoSMoS iterations is depicted, with the first 20 sampling points

CoSMoS iteration, all belonging sampling points are evaluated in parallel via a simple job threading.

2.4 Methodological Aspects of GROW

The GRadient-based Optimization Workflow (GROW) [58] explicitly considers the euclidean norm for the loss function in Eq. (1). GROW is a collection of gradient-based numerical optimization algorithms (e.g., steepest descent, conjugate gradients, and trust region) combined with an efficient Armijo step length control. The latter prevents GROW from both jumping over the minimum and leaving the admissible domain of the force-field parameters. For more details of the algorithms involved in GROW, see Ref. [59].

The gradient at an iteration $x \in \Omega$ is given by the partial derivatives

$$\frac{\partial F}{\partial x_j}(x) = -2\sum_{i=1}^{n} w_i \frac{f_i^{\exp} - f_i^{\text{sim}}(x)}{(f_i^{\exp})^2}\frac{\partial f_i^{\text{sim}}}{\partial x_j}(x),\ j = 1,\ldots,N.$$

The partial derivatives of the properties are approximated numerically by

$$\frac{\partial f_i^{\text{sim}}}{\partial x_j}(x) = \frac{f_i^{\text{sim}}(x_1,\ldots,x_j+h,\ldots,x_N) - f_i^{\text{sim}}(x)}{h},\ h > 0, j = 1,\ldots,N.$$

On the one hand, due to the statistical uncertainties on the simulated properties $f_i^{\text{sim}}(x)$, GROW can get stuck in an intermediate local minimum caused by the noise, if the discretization parameter h is chosen too small. On the other hand, if h is too large, the estimations of the gradient might be incorrect. Hence, a good compromise has to be found, and the choice of h is problem-dependent and thus difficult to optimize in practice. However, GROW turned out to be very successful for the parameterization of force fields in many applications [55, 60–62]. For more algorithmic details concerning GROW, see reference [58].

Local optimization procedures always start with an initial guess $x^0 \in \Omega$, which must be situated in the sphere of influence of the minimum. By evaluating the loss function, the simulated properties are compared with the experimental target data. If a specified stopping criterion is fulfilled, the parameters are final and the workflow ends. Otherwise, for the current iteration $x^k \in \Omega, \quad k \in \mathbb{N}$, GROW searches for a iteration $x^{k+1} \in \Omega$ with a lower loss function. At each iteration, a gradient has to be calculated, whose components are evaluated in parallel together with the original iteration x^k. Note that the force-field parameters for the gradient components are the same as in x^k except for one component which deviates by h from the original one. Hence, at each iteration, $N + 1$ loss function evaluations are parallelized. The Armijo steps are parallelized as well. For each job, time-consuming molecular simulations are required, and parallelization of these simulations reduces the real computation time significantly. Another approach to reduce computational effort consists in efficient gradient computations, which do not require new function evaluations. This is achieved by computing directional derivatives instead of the partial derivatives so that previously performed loss function evaluations can be used again. The same approach can be applied to Hessians (i.e., for the trust region) method as well [63, 64].

The stopping criterion depends on the specific properties to be fitted. For example, if the density deviates by less than 0.5 % from experiment, the corresponding force field is considered as optimal because the experiment is not more accurate either. The same holds for all other properties. However, the experimental accuracy is much lower for transport properties like diffusion coefficients or viscosity.

2.5 *SpaGrOW as an Enhanced GROW-Alternative*

The Sparse Grid-based Optimization Workflow (SpaGrOW) [65] counteracts the drawbacks of local gradient-based optimization mentioned above. It approximates the loss function near the minimum and filters out the statistical noise by regularization methods using naive elastic nets [66]. In order to reduce the computational effort, this approximation is performed on sparse grids [67], meaning that simulations only have to be performed for sparse grid points. As sparse grids are fully occupied at their boundary, transformations onto the unit hypercube is performed, followed by multiplications of the loss function values with sine functions so that they vanish at the boundary and no simulation has to be performed. Afterward, interpolations from sparse to full grids are performed via a combination technique [68], and the loss function is discretely minimized on the resulting full grids.

The integrated trust region approach [59] makes SpaGrOW an iterative procedure: At each iteration, the loss function is considered on a trust region of a certain size. It must be large enough in order to increase the speed of convergence and to distinguish different loss function values despite the statistical noise, and it must be small enough such that the loss function can be reproduced accurately by the sparse grid interpolations. The discrete minimum of the model on the full grid is compared to the corresponding original loss function value. If both coincide well, then the trust region is increased, if not then it is decreased. Due to the grid-based approach, SpaGrOW is able to find a much more direct path to the minimum than GROW. The practical proof that SpaGrOW is able to outperform gradient-based methods for the present optimization task and all algorithmic details can be found in reference [65].

Note that the loss function evaluations for the different sparse grid points are independent from each other. Hence, they are evaluated in parallel like the gradient components within GROW. Due to its derivative-free approach and due to the fact that it leads more directly to the optimum, SpaGrOW is always preferred to GROW within the funnel workflow. However, one or two steepest descent directions may also be reliable after the CoSMoS's global optimization, leading to faster force-field parameters with a lower loss function value. Moreover, SpaGrOW is not suitable for high-dimensional problems due to the involved smoothing and interpolation procedures, whose computation effort increases exponentially with the dimension.

3 Software Realization

3.1 Wolf$_2$Pack

Wolf$_2$Pack is a software package that uses a series of shell scripts that interlink already existing and specialized software (e.g., for computing QM data, statistical analysis, visualization). It enables researchers to optimize intramolecular parameters by fitting to target QM data (i.e., relative energies and geometries) [35, 36]. The QM theories that are possible for generating target data include HF, B3LYP, MP2, AM1, and PM3, while both basis sets proposed by Pople [69] (e.g., 6-31G (d)) and correlation consistent [70] (e.g., aug-cc-pVDZ) basis sets can be specified to describe the orbital space. Currently, *Amber* force fields are available (i.e., Parm14SB [71], Gaff [72], Glycam06j [52], and Lipid14 [73]), as well as our own force field (ExTrM) that is continually being refined and extended.

Parameters optimization can be done using an algorithm or by hand in an iterative process. Several algorithms already exist for intramolecular parameter optimization [1, 6, 53, 74–82]. Currently, we have integrated the algorithms published in Refs. [78, 79]. However, Wolf$_2$Pack strongly encourages the user to perform the optimization by hand in an iterative manner. Doing so allows the users to explore the parameter space and thus build their intuition of how the parameters influence the resulting curves. With gained experience, one can better decide the importance of specific parameters (e.g., a V_3 term in HC–CT–CT–HC), which ones have little influence on given energy curves. For example, an optimization algorithm may determine nonzero values for torsions V_1, V_2, and V_3, while during a manual adjustment, the user observes that the V_2 has little effect on the resulting fit. In such a case, setting the V_2 to zero should lead to an increase in the parameter transferability over diverse molecules. And due to Wolf$_2$Pack's molecular database, such a transferability test can be done easily.

Within Wolf$_2$Pack, all QM calculations are performed by *GAMESS* [83], while all MM calculations are performed by *AmberTools* [1] (i.e., *Sander*). Partial atomic charges are determined using *R.E.D.* [54]. File format conversions are executed using *OpenBabel* [84] and shell scripts. Statistical analysis and image generation are done using *Ptraj* [1], *R statistical language* [85], and *pymol* [86]. LATEX typesetting language, with the graphics and animate packages sourced, is used to generate PDF documents with embedded images of relative energy curves and animations that display an overlay of the resulting QM and MM geometries of each conformation [87]. These PDF files serve to archive the final data and allow for easy dissemination of the results to other researchers.

3.2 CoSMoS, GROW, and SpaGrOW

CoSMoS, GROW, and SpaGrOW are integrated into a fully modular program structure. The program is implemented in a generic manner such that modules can be easily exchanged. This modular structure allows a developer to easily exchange the optimization algorithm, the optimization problem, the objective function, and the constraints. An interface to a new simulation tool can also be easily implemented. The overall structure is object-oriented and easy to extend. All three tools are written in *python* (version 2.6.6). The program is categorized into the following four layers, whereas the first two layers are related to general optimization problems and the last two are related to the execution of molecular simulations:

- Generic Optimization,
- Force-Field Parameterization,
- Parallel Jobs, and
- Simulation.

As shown in Fig. 5, each layer considers two independent optimization sections: the Solver and the Problem Formulation section. The former regards the optimization algorithm itself, while the latter regards the evaluation of the objective function (i.e., the function to be minimized and the constraints). Within the Generic Optimization layer, there are two abstract upper classes, which are the *OptimizationAlgorithm* and *OptimizationProblem* in the Solver and Problem Formulation sections. These two classes are connected in the sense that the *OptimizationAlgorithm* requires a defined

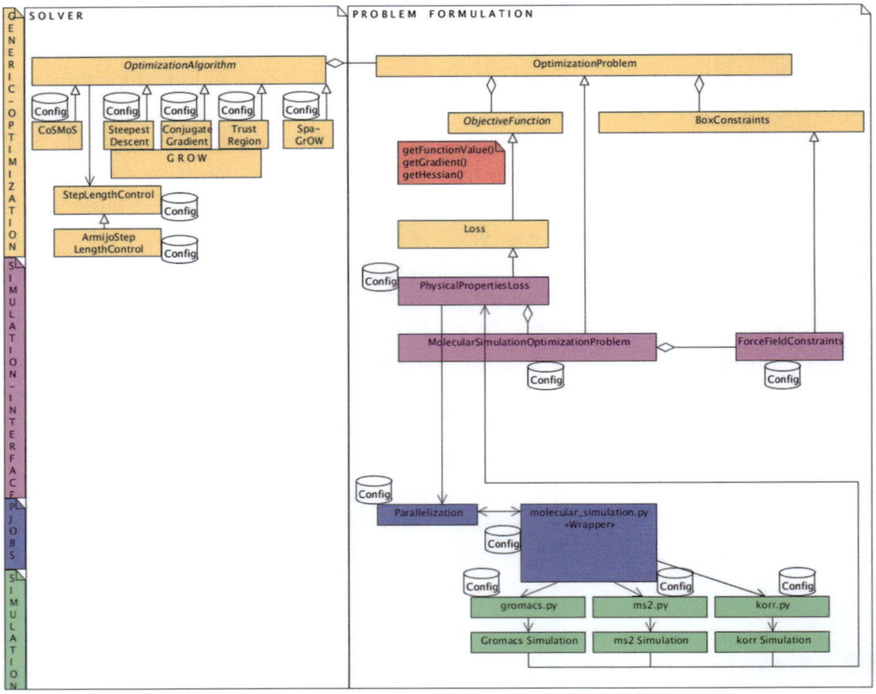

Fig. 5 Generic modular structure of the overall intermolecular optimization toolbox consisting of the abstract layer Generic Optimization and the three specific layers Force-Field (FF) Parameterization, Parallel Jobs (PJOBS), and Simulation. Most of the modules require input parameters, which are defined in the configuration file (i.e., "Config")

problem to solve from *OptimizationProblem*. For *OptimizationProblem,* it is irrelevant which optimization algorithm is used to solve the optimization problem.

Within the Solver section, the class *OptimizationAlgorithm* defines an object of the class *StepLengthControl*, which steers the step length control. The specific class *ArmijoStepLengthControl* is derived from it and can be exchanged by another step length control method other than Armijo. The CoSMoS, GROW, and SpaGrOW algorithms are steered by specific child classes derived from *OptimizationAlgorithm.* GROW itself encompasses the classes *SteepestDescent*, *ConjugateGradients*, and *TrustRegion*.

The optimization problem for *OptimizationAlgorithm* is defined within the Problem Formulation as an objective function to be minimized and box constraints to be met, which are represented by abstract classes *ObjectiveFunction* and *BoxConstraints*. These two classes contain getter and setter functions (e.g., for the function value, the gradient, the Hessian), which have to be overwritten by specific derived child classes in the layer Force-Field Parameterization. A generic loss function class (i.e., *Loss*) is derived from *ObjectiveFunction* implementing a general loss function between calculated and target values (Eq. 1). Its child class *PhysicalPropertiesLoss* steers the molecular simulations that are executed in parallel and collects the simulation results. This module interacts with a wrapper script for the molecular simulation steering calling specific *python* scripts for the desired simulation tools. Currently, interfaces to the simulation tools *Gromacs* [3], *ms2* [88], and *korr* (simulated simulations) [89] are implemented. The molecular simulations can be replaced by so-called *simulated simulations* based on equations of state defining functional dependencies between specific force-field parameters and certain physical observables. This makes it possible to compute physical properties without performing time-consuming molecular simulations (see Refs. [60, 89] for further details).

Finally, an abstract class named *BoxConstraints* is used by *OptimizationProblem* with the specific child class *ForceFieldConstraints* implementing the admissible domain Ω for the force-field parameters. An object of the latter is given to the class *MolecularSimulationOptimizationProblem* derived from the abstract class *OptimizationProblem*. Once the simulation results (i.e., the simulated physical properties) have been calculated, they are given back to the class *PhysicalPropertiesLoss*.

A majority of the modules requires certain input parameters, which have to be defined in a user-written configuration file, and is read by the main python module *main.py*. The configuration file specifies all class objects, modules, and submodules that are desired for optimization process. It also contains important preferences concerning the system (e.g., input/output paths, number of computer cores, batch system), the optimization (e.g., algorithm, step length control, stopping criterion, initial parameters, constraints), and the optimization problem (e.g., objective functions, the loss function's target values). When molecular simulations are performed, all desired properties and parameters of the thermodynamic system have to be defined (e.g., ensemble, temperatures, pressures, physical properties to be fitted, number of molecules, box size, number of MD/MC steps, time step). Hence, the file is divided into three blocks. If more than one substance is considered in the optimization, one block for each substance has to be indicated.

The final output file contains an evaluation in tabular form of all simulation and optimized force-field parameters, the simulated properties along with their actual deviations from the experimental reference data at each temperature, the loss function values, and algorithm-specific information.

The steering of parallel molecular simulations requires special consideration. This is realized by three different modules: the producer, the executer, and the collector. The main function of the *producer*, illustrated in Fig. 6, is to generate all configuration files for the molecular simulations. In order to generate transferable force fields, a variation level was added to the producer. This allows researchers to vary their optimization jobs by the force-field parameters, number of ensembles, temperatures, and molecular models (i.e., different substances). Before running the producer, the user must define all model systems with their properties in the initial configuration file, which contains several sections for each system. The relevant

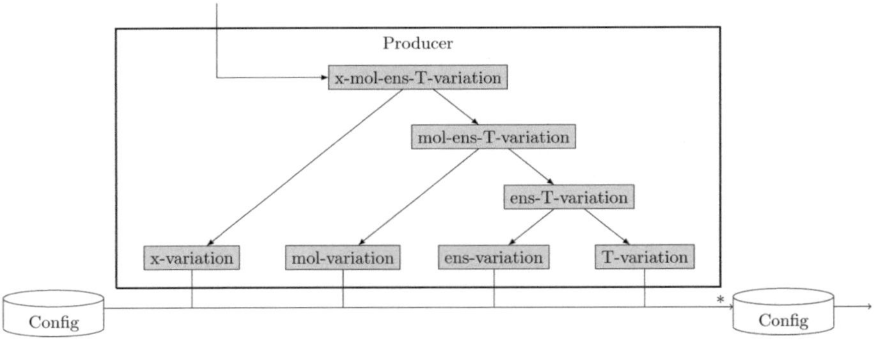

Fig. 6 Illustration of the producer module comprising the *x-variation*, *mol-variation*, *ens-variation*, and *T-variation* scripts

properties for the producer are the force-field parameters, substances, ensembles, and temperatures.

Generally, all necessary configuration files are realized in the following manner. First, the *x-mol-ens-T-variation* script is started, which calls the *x-variation* script. This script then reads the initial configuration file and generates subdirectories that contain new configuration files with the new force-field parameters as varied by the optimization algorithm. Second, the *mol-ens-T-variation* script calls the *mol-variation* script, which varies the new configuration files with respect to different substances and stores them in new subdirectories. Third, the *ens-T-variation* script calls the *ens-variation* script. This script then reads the new configuration files and varies the ensembles as well. The new files are stored into subdirectories. Finally, the *T-variation* script is called, varying the temperature and storing the new configuration files into a new subdirectory. In summary, the producer generates a four-level subdirectory structure with varied configuration files, as exemplified in Fig. 7, according to the following pattern: force-field parameters–substances–ensembles–temperatures.

After this procedure, the *executer* starts the parallel molecular simulations based on the set of configuration files. After completion, the executer reports the status and results of all simulations to the *collector*. The latter collects the simulation results of each single molecular simulation being stored in the leaf subdirectory level. The main idea is that the collector runs through all result folders, collects the simulated physical properties, and stores them together in a result file within the highest directory level. Afterward, the result file is used for the evaluation of the loss function.

Fig. 7 Illustration of the four-level subdirectory structure that is generated by the producer module. A unique configuration file is stored in all subdirectories

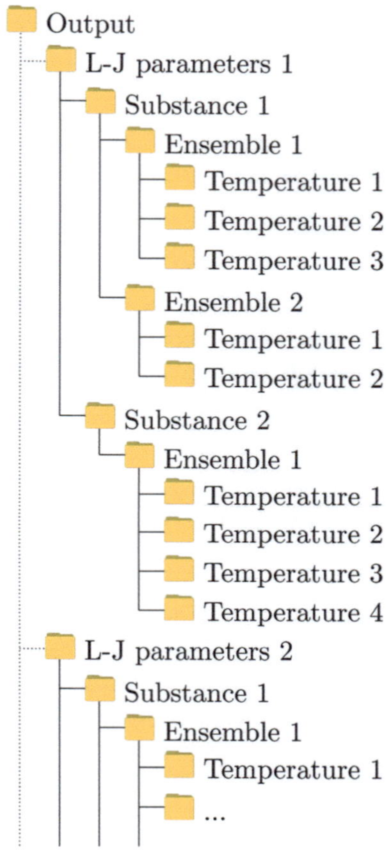

4 Interlinking Aspects of Bonded and Nonbonded Parameter Optimization

It is well known that bonded and nonbonded parameters are coupled to each other. For a given set of nonbonded parameters, there will be an optimal set of bonded parameters and vice versa. This implies that through a successive iteration of bonded and nonbonded parameter optimization, a self-consistent force field should be achieved. Figure 8 shows the interaction between intramolecular and intermolecular parameter optimization tools. Often, an initial set of Lennard–Jones parameters is chosen based on existing force fields and atom types. One then optimizes the bonded parameters using Wolf$_2$Pack. The resulting parameters are then transferred to the intermolecular optimization tools, which optimizes the nonbonded parameters. Depending on the algorithm used, the transferred Lennard–Jones parameters are used as an initial guess (i.e., GROW and SpaGrOW) or they are discarded (i.e., CoSMoS). Once new nonbonded parameters are generated, they

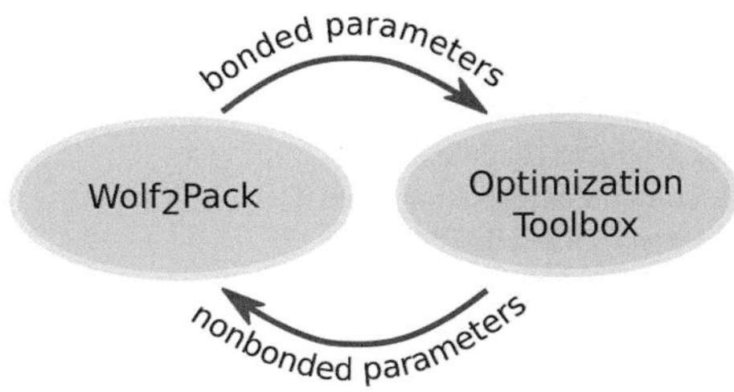

Fig. 8 Interaction between intramolecular (i.e., $Wolf_2Pack$) and the intermolecular parameter optimization tools (i.e., CoSMoS, GROW, SpaGrOW)

are then transferred back to $Wolf_2Pack$, and the cycle is repeated until all investigated parameters converge. Currently, we are improving our understanding of the sensitivity of this global optimization routine by performing it on selected saturated hydrocarbons (e.g., octane).

5 Future Work: Methods and Applications

In addition to researching how to best realize the bonded–nonbonded optimization cycle described in the last section, we are currently working toward the inclusion of solution-phase models (e.g., pure solvent PBC box, ionic liquid PBC boxes) into $Wolf_2Pack$'s database. Experimentally known condense-phase observables (e.g., density, enthalpy of vaporization) will also be included into the database. These models and target experimental values will be accessible to CoSMoS, GROW, and SpaGrOW. This will allow future users to have a common access point and starting models for nonbonded parameter optimization. Once this is realized, the next step will be to extend $Wolf_2Pack$'s online portal to include these condensed-phase models and our nonbonded optimization algorithms, thus unifying our bonded and nonbonded software packages.

With regard to application, we will apply our tools to optimize a force field specific for fluorinated alcohols. Fluorinated alcohols are highly relevant in industrial applications (e.g., as solvents used in chemical separation processes). Their attractiveness is that they can be extracted from the reaction medium and be reused, which makes them both environmentally friendly and economically attractive [90]. The challenge in optimizing such a force field arises from the lack of experimental data and lacks previously published parameters that can be used as an initial input [91–93]. The goal will be to fit both vapor–liquid equilibrium data (e.g., saturated liquid density, vapor pressure) and transport properties (e.g., diffusion coefficients) simultaneously and at different temperatures. Hence, not only parallelization over different substances but also over different ensembles and temperatures are required.

Furthermore, a new force field for carbon dioxide will be developed that reproduces bulk densities, vapor–liquid equilibrium data, and overcritical transport properties (e.g., diffusion coefficients and viscosities) simultaneously. New force fields for alkaline earth salts, including a transferable parameters, are about to be published.

6 Conclusion

In this work, the conception and implementation of recently developed modular program packages applied for force-field parameterizations was described in detail. Intramolecular parameters (i.e., bond length, angles, and torsions) are obtained using the software package $Wolf_2Pack$. Intermolecular parameters, especially Lennard–Jones parameters, are computed via a new set of software tools, implementing a so-called funnel workflow combing global and local optimization procedures. The global metamodeling package CoSMoS is combined with gradient-based (GROW) or derivative-free methods (SpaGrOW). The derivative-free method, based on smoothing procedures and sparse grid interpolation, tends to be much more efficient near the global optimum. The mathematical optimization problem is formulated through the minimization of a loss function between simulated physical properties and experimental reference data. It was shown how the individual software is interlinked with each other within the overall optimization package. These tools form the basis for user-friendly and highly efficient parallelized force-field parameterizations. Finally, several applications are planed in order to obtain industrially relevant force fields (i.e., for solution-phase models, ionic liquids, fluorinated alcohols, alkaline earth salts, and overcritical CO_2).

References

1. Case, D.A., Babin, V., Berryman, J.T., Betz, R.M., Cai, Q., Cerutti, D.S., Cheatham III, T.E., Darden, T.A., Duke, R.E., Gohlke, H., Goetz, A.W., Gusarov, S. Homeyer, N., Janowski, P., Kaus, J., Kolossváry, I., Kovalenko, A., Lee, T.S., LeGrand, S., Lucko, T., Luo, R., Madej, B., Merz, K.M., Paesani, F., Roe, D.R., Roitberg, A., Sagui, C., Salomon-Ferrer, R., Seabra, G., Simmerling, C.L., Smith, W., Swails, J., Walker, R.C., Wang, J., Wolf, R.M., Wu, X., Kollmann, P.A.: AMBER 14. http://ambermd.org. University of California, San Francisco (2014)
2. Brooks, B.R., Brooks III, C.L., Mackerell, A.D., Nilsson, L., Petrella, R.J., Roux, B., Won, Y., Archontis, G., Bartels, C., Caflisch, S.B.A., Caves, L., Cui, Q., Dinner, A.R., Feig, M., Fischer, S., Gao, J., Hodoscek, M., Im, W., Kuczera, K., Lazaridis, T., Ma, J., Ovchinnikov, V., Paci, E., Pastor, R.W., Post, C.B., Pu, J.Z., Schaefer, M., Tidor, B., Venable, R.M., Woodcock, H.L., Wu, X., Yang, W., York, D.M., Karplus, M.: Charmm: the biomolecular simulation program. J. Comp. Chem. **30**, 1545–1615 (2009)
3. Hess, B., van der Spoel, D., Lindahl, E.: Gromacs user manual 4.5.4. http://www.gromacs.org/Documentation/Manual/manual-4.5.4.pdf (2010)
4. Plimpton, S.: Fast parallel algorithms for short-range molecular dynamics. J. Comp. Phys. **117**, 1–19 (1995)

5. Gil, Y., Deelman, E., Ellisman, M., Fahringer, T., Fox, G., Gannon, D., Goble, C., Livny, M., Moreau, L., Myers, J.: Examining the challenges of scientific workflows. Computer **40**, 24–32 (2007)
6. Waldher, B., Kuta, J., Chen, S., Henson, N., Clark, A.E.: ForceFit: a code to fit classical force fields to quantum mechanical potential energy surfaces. J. Comp. Chem. **12**, 2307–2316 (2010)
7. Highly optimized object-oriented many-particle dynamics—blue edition. http://codeblue.umich.edu/hoomd-blue/ (2011)
8. Halverson, J.D., Brandes, T., Lenz, O., Arnold, A., Bevc, S., Starchenko, V., Kremer, K., Stuehn, T., Reith, D.: ESPResSo++: a modern multiscale simulation package for soft matter systems. Comput. Phys. Commun. **184**, 1129–1149 (2013)
9. Karimi-Varzaneh, H., Qian, H., Chen, X., Carbone, P., Müller-Plathe, F.: Ibisco: a molecular dynamics simulation package for coarse-grained simulation. J. Comp. Chem. **32**, 1475–1487 (2011)
10. Singer, S.J., Nicolson, G.L.: The fluid mosaic model of the structure of cell membranes. Science **175**, 720–731 (1972)
11. Zhou, Y., Stell, G.: Chemical association in simple models of molecular and ionic fluids II. Thermodynamic properties. J. Chem. Phys. **96**, 1504–1506 (1992)
12. Siepmann, J.I., Karaborni, S., Smit, B.: Simulating the critical behaviour of complex fluids. Nature **365**, 330–332 (1993)
13. O'Connell, S.T., Thompson, P.A.: Molecular dynamics-continuum hybrid computations: a tool for studying complex fluid flow. Phys. Rev. E **52**, 5792–5795 (1995)
14. Kolafa, J., Nezbeda, I., Lisal, M.: Effect of short- and long-range forces on the properties of fluids. III. dipolar and quadrupolar fluids. Mol. Phys. **99**, 1751–1764 (2001)
15. Valiullin, R., Naumov, S., Galvosas, P., Kärger, J., Woo, H.-J., Porcheron, F., Monson, P.A.: Exploration of molecular dynamics during transient sorption of fluids in mesoporous materials. Nature **443**, 965–968 (2006)
16. Batra, I.P., Bennett, B.I., Herman, F.: Simple molecular model for crystalline tetrathiofulvalene-tetracyanoquinodimethane (TTF-TCNQ). Phys. Rev. B **11**, 4927–4934 (1975)
17. Fehlner, T.P.: Molecular models of solid state metal boride structure. J. Solid State Chem. **154**, 110–113 (2000)
18. Della, C.N., Dongwei, S.: Mechanical properties of carbon nanotubes reinforced ultra high molecular weight polyethylene. Solid State Phenom. **136**, 45–49 (2008)
19. Lin, S.-T., Blanco, M., Goddard III, W.A.: The two-phase model for calculating thermodynamic properties of liquids from molecular dynamics: validation for the phase diagram of Lennard-Jones fluids. J. Chem. Phys. **119**, 11792–11805 (2003)
20. Bien, D.E., Chiriac, V.A.: A novel molecular approach to modeling phase change in micro-fluidic systems. In: Proceedings of the 9th Intersociety Conference on Thermal and Thermomechanical Phenomena in Electronic Systems, pp. 598–604. IEEE, New Jersey (2004)
21. Vrabec, J., Gross, J.: Vapor–liquid equilibria simulation and an equation of state contribution for dipole-quadrupole interactions. J. Phys. Chem. B **112**, 51–60 (2008)
22. Levitt, M., Warshel, A.: Computer simulation of protein folding. Nature **253**, 694–698 (1975)
23. Gsponer, J., Caflisch, A.: Molecular dynamics simulations of protein folding from the transition state. In: Fersth, A. (ed.) Proceedings of the National Academy of Sciences (PNAS), vol. 99, pp. 6719–6724. Washington (2002)
24. Snow, C.D., Sorin, E.J., Rhee, Y.M., Pandel, V.S.: How well can simulation predict protein folding kinetics and thermodynamics? Annu. Rev. Biophys. Biomol. Struct. **34**, 43–69 (2005)
25. Hodgkin, A.L., Huxley, A.F.: A quantitative description of membrane current and its application to conduction and excitation in nerve. J. Physiol. **117**, 500–544 (1952)
26. Barkla, B.J., Pantoja, O.: Physiology of ion transport across the tonoplast of higher plants. Annu. Rev. Plant Physiol. Plant Mol. Biol. **47**, 159–184 (1996)
27. Müller-Plathe, F., Reith, D.: Cause and effect reversed in non-equilibrium molecular dynamics: an easy route to transport coefficients. Comput. Theor. Polymer Sci. **9**, 203–209 (1999)

28. Bordat, P., Reith, D., Müller-Plathe, F.: The influence of interaction details on the thermal diffusion in binary Lennard-Jones liquids. J. Chem. Phys. **115**, 8978–8982 (2001)
29. Guevara-Carrion, G., Nieto-Draghi, C., Vrabec, J., Hasse, H.: Prediction of transport properties by molecular simulation: methanol and ethanol and their mixture. J. Phys. Chem. B **112**, 16664–16674 (2008)
30. Grest, G.S., Kremer, K.: Molecular dynamics simulation for polymers in the presence of a heat bath. Phys. Rev. A **33**, 3628–3631 (1986)
31. Müller-Plathe, F.: Permeation of polymers—a computational approach. Acta Polymer. **45**, 259–293 (1994)
32. Binder, K.: Monte Carlo and molecular dynamics simulations in polymer science. Oxford University Press, Oxford (1995)
33. Kremer, K., Müller-Plathe, F.: Multiscale simulation in polymer science. Mol. Sim. **28**, 729–750 (2002)
34. Praprotnik, M., Junghans, C., Delle Site, L., Kremer, K.: Simulation approaches to soft matter: generic statistical properties vs. chemical details. Comput. Phys. Commun. **179**, 51–60 (2008)
35. Reith, D., Kirschner, K.N.: A modern workflow for force field development—bridging quantum mechanics and atomistic computational models. Comput. Phys. Commun. **182**, 2184–2191 (2011)
36. Krämer-Fuhrmann, O., Neisius, J., Gehlen, N., Reith, D., Kirschner, K.N.: Wolf$_2$Pack – Portal based atomistic force field development. J. Chem. Inf. Mod. **53**, 802–808 (2013)
37. Vainio, M.J., Johnson, M.S.: Generating conformer ensembles using a multiobjective genetic algorithm. J. Chem. Inf. Mod. **47**, 2462–2474 (2007)
38. Tasi, G., Mizukami, F., Csontos, J., Gyõrffy, W., Pálinkó, I.: Quantum algebraic–combinatoric study of the conformational properties of *n*-alkanes. II. J. Math. Chem. **27**, 191–199 (2000)
39. Jorgensen, W.L., Madura, J.D., Swensen, C.J.: Optimized intermolecular potential functions for liquid hydrocarbons. J. Am. Chem. Soc. **106**, 6638–6646 (1984)
40. Martin, M.G., Siepmann, J.I.: Transferable potentials for phase equilibria. 1. United-atom description of *n*-alkanes. J. Phys. Chem. B **102**, 2569–2577 (1998)
41. Ungerer, P., Beauvais, C., Delhommelle, J., Boutin, A., Rousseau, B., Fuchs, A.H.: Optimization of the anisotropic united atoms intermolecular potential for *n*-alkanes. J. Phys. Chem. **112**, 5499–5510 (2000)
42. Bourasseau, E., Haboudou, M., Boutin, A., Fuchs, A.H., Ungerer, P.: New optimization method for intermolecular potentials: optimization of a new anisotropic united atoms potential for olefins: prediction of equilibrium properties. J. Chem. Phys. **118**, 3020–3035 (2003)
43. Stoll, J., Vrabec, J., Hasse, H.: A set of molecular models for carbon monoxide and halogenated hydrocarbons. J. Chem. Phys. **119**, 11396–11407 (2003)
44. Reith, D., Pütz, M., Müller-Plathe, F.: Deriving effective mesoscale potentials from atomistic simulations. J. Comp. Chem. **24**, 1624–1636 (2003)
45. Oostenbrink, C., Villa, A., Mark, A.E., van Gunsteren, W.F.: A biomolecular force field based on the free enthalpy of hydration and solvation: the GROMOS force-field parameter sets 53A5 and 53A6. J. Comp. Chem. **25**, 1656–1676 (2004)
46. Sun, H.: Prediction of fluid densities using automatically derived VDW parameters. Fluid Phase Eq. **217**, 59–76 (2004)
47. Eckl, B., Vrabec, J., Hasse, H.: On the application of force fields for predicting a wide variety of properties: ethylene oxide as an example. Fluid Phase Eq. **274**, 16–26 (2008)
48. Cacelli, I., Cimoli, A., Livotto, P.R., Prampolini, G.: An automated approach for the parameterization of accurate intermolecular force-fields: pyridine as a case study. J. Comp. Chem. **33**, 1055–1067 (2012)
49. Ucyigitler, S., Camurdan, M.C., Elliott, J.R.: Optimization of transferable site–site potentials using a combination of stochastic and gradient search algorithms. Ind. Eng. Chem. Res. **51**, 6219–6231 (2012)
50. Eckelsbach, S., Janzen, T., Köster, A., Mirshnichenko, S., Muñoz Muñoz, Y.M., Vrabec, J.: Molecular models for cyclic alcanes and ethyl acetate as well as surface tension data from molecular simulation. In: Nagel, W.E., Kröner, D.E., Resch, M.M. (eds.) High Performance

Computing in Science and Engineering '14, Transactions of the High Performance Computing Center, HLRS, Stuttgart (2014), pp. 645–659. Springer, Berlin (2015)
51. Muñoz Muñoz, Y.M., Guevara-Carrion, G., Llano-Restrepo, M., Vrabec, J.: Lennard–Jones force field parameters for cyclic alkanes from cyclopropane to cyclohexane. Fluid Phase Eq. **404**, 150–160 (2015)
52. Kirschner, K.N., Yongye, A.B., Tschampel, S.M., Gonzalez-Outeirino, J., Daniels, C.R., Foley, B.L., Woods, R.J.: GLYCAM06: a generalizable biomolecular force field. Carbohydrates. J. Comp. Chem. **29**, 622–655 (2008)
53. Faller, R., Schmitz, H., Biermann, O., Müller-Plathe, F.: Automatic parameterization of force fields for liquids by simplex optimization. J. Comp. Chem. **20**, 1009–1017 (1999)
54. Dupradeau, F.-Y., Pigache, A., Zaffran, T., Savineau, C., Lelong, R., Grivel, N., Lelong, D., Rosanski, W., Cieplak, P.: The R.E.D. tools: advances in RESP and ESP charge derivation and force field library building. Phys. Chem. Chem. Phys. **12**, 7821–7839 (2010)
55. Hülsmann, M.: Effiziente und neuartige Verfahren zur Optimierung von Kraftfeldparametern bei atomistischen Molekularen Simulationen kondensierter Materie. In: Fraunhofer SCAI (ed.) Fraunhofer-Verlag, Ph.D. thesis, University of Cologne, Germany (2012)
56. Krämer, A., Hülsmann, M., Köddermann, T., Reith, D.: Automated parameterization of intermolecular pair potentials using global optimization techniques. Comput. Phys. Commun. **185**, 3228–3239 (2014)
57. Regis, R., Shoemaker, C.: Constrained global optimization of expensive black box functions using radial basis functions. J. Glob. Opt. **31**, 153–171 (2005)
58. Hülsmann, M., Köddermann, T., Vrabec, J., Reith, D.: GROW: A gradient-based optimization workflow for the automated development of molecular models. Comput. Phys. Commun. **181**, 499–513 (2010)
59. Nocedal, J., Wright, S.J.: Numerical Optimization. Springer, New York (1999)
60. Hülsmann, M., Vrabec, J., Maaß, A., Reith, D.: Assessment of numerical optimization algorithms for the development of molecular models. Comput. Phys. Commun. **181**, 887–905 (2010)
61. Hülsmann, M., Müller, T.J., Köddermann, T., Reith, D.: Automated force field optimization of small molecules using a gradient-based workflow package. Mol. Sim. **36**, 1182–1196 (2011)
62. Köddermann, T., Kirschner, K.N., Vrabec, J., Hülsmann, M., Reith, D.: Liquid-liquid equilibria of dipropylene glycol dimethyl ether and water by molecular dynamics. Fluid Phase Eq. **310**, 25–31 (2011)
63. Hülsmann, M., Kopp, S., Huber, M., Reith, D.: Efficient gradient and Hessian calculations for numerical optimization algorithms applied to molecular simulations. In: Proceedings of the International Conference on Mathematical Modeling in Physical Sciences (IC-MSQUARE), Budapest, Hungary (2012), IOP Publishing, Journal of Physics: Conference Series 410, 012007 (2013)
64. Hülsmann, M., Kopp, S., Huber, M., Reith, D.: Utilization of efficient gradient and Hessian computations in the force field optimization process of molecular simulations. Comput. Sci. Disc. **6**, 015005 (2013)
65. Hülsmann, M., Reith, D.: SpaGrOW—a derivative-free optimization scheme for intermolecular force field parameters based on sparse grids methods. Entropy **15**, 3640–3687 (2013)
66. Zou, H., Hastie, T.: Regularization and variable selection via the elastic net. J. Roy. Stat. Soc. Ser. B **67**, 301–320 (2005)
67. Smolyak, S.A.: Quadrature and interpolation formulas for tensor products of certain classes of functions. Sov. Math. Doklady **4**, 240–243 (1963)
68. Griebel, M., Schneider, M., Zenger, C.: A combination technique for the solution of sparse grid problems. Technical Report, Institute for Computer Science, Technical University of Munich, Germany (1990)
69. Ditchfield, R., Hehre, W.J., Pople, J.A.: Self consistent molecular orbital methods. IX. An extended Gaussian type basis for molecular orbital studies of organic molecules. J. Chem. Phys. **54**, 724–728 (1971)

70. Dunning, T.H.: Gaussian basis sets for use in correlated molecular calculations. I. The atoms boron through neon and hydrogen. J. Chem. Phys. **90**, 1007–1023 (1989)
71. Maier, J.A., Martinez, C., Kasavajhala, K., Wickstrom, L., Hauser, K.E., Simmerling, C.: ff14SB: improving the accuracy of protein side chain and backbone parameters from ff99SB. J. Chem. Theory Comput. **11**, 3696–3713 (2015)
72. Wang, J., Wolf, R.M., Caldwell, J.W., Kollman, P.A., Case, D.A.: Development and testing of a general amber force field. J. Comput. Chem. **25**, 1157–1174 (2004)
73. Dickson, C.J., Madej, B.D., Skjevik, Å.A., Betz, R.M., Teigen, K., Gould, I.A., Walker, R.C.: Lipid14: the amber lipid force field. J. Chem. Theory Comput. **10**, 865–879 (2014)
74. Wang, J.M., Kollman, P.A.: Automatic parameterization of force field by systematic search and genetic algorithms. J. Comp. Chem. **22**, 1219–1228 (2001)
75. Vaiana, A.C., Cournia, Z., Costescu, I.B., Smith, J.C.: AFMM: a molecular mechanics force field vibrational parametrization program. Comput. Phys. Commun. **167**, 34–42 (2005)
76. Guvench, O., MacKerell Jr, A.D.: Automated conformational energy fitting for force field development. J. Mol. Model. **14**, 667–679 (2008)
77. Mayne, C.G., Saam, J., Schulten, K., Tajkhorshid, E., Gumbart, J.C.: Rapid parameterization of small molecules using the force field toolkit. J. Comp. Chem. **32**, 2757–2770 (2013)
78. Hopkins, C.W., Roitberg, A.E.: Fitting of dihedral terms in classical force fields as an analytic linear least-squares problem. J. Chem. Inf. Mod. **54**, 1978–1986 (2014)
79. Burger, S.K., Ayers, P.W., Schofield, J.: Efficient parameterization of torsional terms for force fields. J. Comp. Chem. **35**, 1438–1445 (2014)
80. Betz, R.M., Walker, R.C.: Paramfit: automated optimization of force field parameters for molecular dynamics simulations. J. Comp. Chem. **36**, 79–87 (2015)
81. Vanommeslaeghe, K., Mingjun, Y., MacKerell, A.D.: Robustness in the fitting of molecular mechanics parameters. J. Comp. Chem. **36**, 1083–1101 (2015)
82. Vanduyfhuys, L., Vandenbrande, S., Verstraelen, T., Schmid, R., Waroquier, M., Van Speybroeck, V.: QuickFF: a program for a quick and easy derivation of force fields for metal-organic frameworks from ab initio input. J. Comp. Chem. **36**, 1015–1027 (2015)
83. Gordon, M.D., Schmidt, M.W.: Advances in electronic structure theory: GAMESS a decade later. In: Gordon, M.S., Schmidt, W., Dykstra, C.E. (eds.) Theory and Applications of Computational Chemistry: The First Forty Years, pp. 1167–1189. Elsevier Amsterdam Boston (2005)
84. O'Boyle, N., Banck, M., James, C., Morley, C., Vandermeersch, T., Hutchison, G.: Open babel: an open chemical toolbox. J. Cheminf. **3**, 33 (2011)
85. R: A language and environment for statistical computing. manual. http://www.R-project.org. The R Foundation for Statistical Computing, Vienna, Austria (2009)
86. PyMOL(TM) Molecular Graphics System, Version 1.6.0.0. http://pymol.org && http://sourceforge.net/projects/pymol/ (2009)
87. The LaTeX Project. http://latex-project.org/
88. Deublein, S., Eckl, B., Stoll, J., Lishchuk, S.V., Guevara-Carrion, G., Glass, C.W., Merker, T., Bernreuther, M., Hasse, H., Vrabec, J.: ms2: a molecular simulation tool for thermodynamic properties. Comput. Phys. Commun. **182**, 2350–2367 (2011)
89. Stoll, J., Vrabec, J., Hasse, H., Fischer, J.: Comprehensive study of the vapour–liquid equilibria of the pure two–centre Lennard-Jones plus point quadrupole fluid. Fluid Phase Eq. **179**, 339–362 (2001)
90. Bégué, J.-P., Bonnet-Delpon, D., Crousse, B.: Fluorinated alcohols: anew medium for selective and clean reaction. Synlett, 18–29 (2004)
91. Rochester, C.H., Symonds, J.R.: Densities of solutions of four fluoralcohols in water. J. Fluorine Chem. **4**, 141–148 (1974)
92. Gross, T., Karger, N., Price, W.E.: p, T dependence of self-diffusion in 2-fluoroethanol, 2,2 difluoroethanol and 2,2,2-trifluoroethanol. J. Mol. L. **75**, 159–168 (1998)
93. Meeks, A.C., Goldfarb, I.J.: Vapor pressure of fluoroalcohols. J. Chem. Eng. Data **12**, 196 (1967)

A Hierarchical, Component Based Approach to Screening Properties of Soft Matter

Christoph Klein, János Sallai, Trevor J. Jones, Christopher R. Iacovella, Clare McCabe and Peter T. Cummings

Abstract In prior work, Sallai, et al. introduced the concept and algorithms of building molecular topologies through the use of a hierarchical data structure and the use of an affine coordinate transformation to connect molecular components. In this work, we expand upon the original concept and present a refined version of this software, termed `mBuild`, which is a general tool for constructing arbitrarily complex input configurations for molecular simulation in a programmatic fashion. Basic molecular components are connected using an equivalence operator which reduces and often removes the need for users to explicitly rotate and translate components as they assemble systems. Additionally, the programmatic nature of this approach and integration with the scientific Python ecosystem seamlessly exposes high-level variables that users can tune to alter the chemical composition of their systems, such as mixtures of polymers of different chain lengths and surface patterning. Leveraging these features, we demonstrate how `mBuild` serves as a stepping stone towards screening and performing optimizations in chemical

C. Klein (✉) · T.J. Jones · C.R. Iacovella · C. McCabe · P.T. Cummings
Department of Chemical and Biomolecular Engineering, Vanderbilt University, Nashville, TN 37235, USA
e-mail: christoph.klein@vanderbilt.edu

P.T. Cummings
e-mail: peter.cummings@vanderbilt.edu

J. Sallai
Institute for Software Integrated Systems, Vanderbilt University, Nashville, TN 37235, USA

C. Klein · C.R. Iacovella · C. McCabe · P.T. Cummings
Vanderbilt Multiscale Modeling and Simulation (MuMS) Facility, Vanderbilt University, Nashville, TN 37235, USA

C. McCabe
Department of Chemistry, Vanderbilt University, Nashville, TN 37235, USA

R.Q. Snurr et al. (eds.), *Foundations of Molecular Modeling and Simulation*, Molecular Modeling and Simulation, DOI 10.1007/978-981-10-1128-3_5

parameter space of complex materials by performing automated screening studies of monolayer systems as a function of graft type, degree of polymerization, and surface density.

Keywords Molecular dynamics · Software · System construction

1 Introduction

The biophysics simulation community has put considerable effort into creating tools and databases for building and parameterizing biological molecules with minimal effort, e.g. the Protein Data Bank [1], VMD [2], AmberTools [3], the Omnia suite [4]. Such toolchains allow researchers to generate input files for complex structures, such as proteins and DNA, that can run on most molecular dynamics simulation engines with little to no manual intervention. However, while the biophysics community's tools provide excellent functionality for biological system setup, they do not allow one to easily generate arbitrary structures found outside the biophysics community. For example, surface bound brushes or tethered nanoparticles, which often feature semi-infinite substrates and/or irregular surface bonding sites, require a less specialized approach. These systems may not be regular and thus defining a small unit cell and replicating it is not always possible. Additionally, many tools are tied to a specific simulation environment [3] or are operated via a custom language that complicates integration with a broader scientific ecosystem of tools for performing tasks not specific to the domain of molecular simulation, such as statistical analysis and visualization.

In prior work [5], we introduced the preliminary concepts underpinning `mBuild`'s functionality. Since then, `mBuild` has evolved into a Python package designed to simplify the construction of complex, regular and irregular structures and topologies as well as integrate seamlessly with the Python scientific stack and more recently developed Python tools in the area of molecular simulation [6–10]. `mBuild` adopts a hierarchical approach to system construction that relies on equivalence relations to connect chemical building blocks (components). Every component can recursively contain particles and other components to generate arbitrary, hierarchical structures where every particle represents a leaf in the hierarchy. Low-level components, such as an alkyl group or a monomer, can be hand-drawn using software like Avogadro [11] and then connected using an equivalence operator which matches defined attachment sites between two components—the operator forces two sets of points in space to overlap thus translating and rotating components into the desired positions. This approach minimizes and often even eliminates the need for users to explicitly translate or rotate components while constructing initial configurations—users simply specify which components should be connected. Additionally, the hierarchical nature of this approach allows for complex families of chemical structures to be encapsulated in a single component class which exposes user defined, tunable parameters that adjust the structural properties of the system (e.g. chain length, surface coverage). By providing a more natural avenue to express such structures, where the requirement for mental visualization of spatial arrangements is minimized, `mBuild` provides a stepping stone towards the goals outlined by the Materials Genome Initiative [12], by enabling screening of and optimizations in chemical parameter space of complex, soft-materials.

Here, we provide an overview of the algorithms associated with `mBuild` including several recent improvements, and demonstrate its use as a means for automating screening of soft matter systems. We illustrate the construction of basic components, how they can be connected programmatically into complex chemical systems, and finally showcase this functionality by generating and performing parameter sweeping simulations of an ensemble of monolayers constructed of alkanes and polyethylene glycol (PEG) where, through the functionality of `mBuild`, we trivially vary surface density, patterning and chain length in an automated, programmatic way.

2 Software Concept

While the basic concepts and algorithms underlying `mBuild` were outlined in Ref. [5] additional refinement and development has been undertaken, as reported here, in particular to simplify and increase the generality of the data structure and provide enhancements with regards to connecting individual components via equivalence transforms. The primary building blocks of an `mBuild` hierarchy are `Compounds`; every user-created component inherits from this class. Each `Compound` can contain an arbitrary amount of other `Compounds`, allowing for systems to be flexibly built in a hierarchical manner. The programmatic connection of `Compounds` in three dimensional space is facilitated by an equivalence transform. This concept is formalized and implemented via the `Port` class which defines connection sites and orientation. These are each discussed below.

2.1 *Data Structure*

The hierarchical data structure of `mBuild` is composed of `Compounds`. `Compounds` maintain an ordered set of children which are other `Compounds`. `Compounds` at the bottom of an `mBuild` hierarchy, i.e., the leafs of the tree, are referred to as `Particles` and can be instantiated as, for example, `lj = mb.Particle(name='lennard-jonesium')`. Note however, that this merely serves to illustrate that this `Compound` is at the bottom of the hierarchy; `Particle` is an alias for `Compound` which can be used to clarify the intended role of an object you are creating.

Every `mBuild` hierarchy also maintains a network of bonds between its `Particles` in the form of a graph as provided by the `NetworkX` package [13]. This graph is maintained by the root (top level component) of the given hierarchy. When two `Compounds` with bonds are added together, their bond graphs are composed.

Additionally, `Compounds` have built-in support for copying and deep copying `Compound` hierarchies, enumerating particles or bonds in the hierarchy, proximity based searches, visualization, I/O operations, and a number of other convenience methods that enable complex topologies to be constructed with little user effort.

2.2 *Equivalence Transforms*

When connecting components in 3D space, their relative orientations must be specified. In `mBuild`, this is accomplished via an equivalence transform. The equivalence operator described here declares points in a component's local coordinate system to be equivalent to points in another component's coordinate system. Using these point pairs, it is possible to compute a rigid transformation, specifically an affine coordinate transformation conserving scaling and orientation (chirality), that, when applied to one component, will transform its designated points to the other component's respective points. Specifying four or more pairs of non-coplanar points is sufficient to compute an unambiguous transformation matrix in 3D space.

Using a rigid transformation F, one can map a point vector υ to its image $F(\upsilon)$ in a different coordinate system. This operation can be expressed as a multiplication by a rotation matrix $R \in \mathbb{R}^{3\times 3}$ and a translation with vector $t \in \mathbb{R}^{3\times 1}$.

$$F(\upsilon) = R\upsilon + t \tag{1}$$

R and t can be solved for using the singular value decomposition to get the pseudoinverse given four or more points $P_i(x_i, y_i, z_i)$ and their images $P_i^{'}(x_i^{'}, y_i^{'}, z_i^{'})$ in the target 3-dimensional coordinate system:

$$\begin{bmatrix} x_1^{'} & x_2^{'} & \dots & x_n^{'} \\ y_1^{'} & y_2^{'} & \dots & y_n^{'} \\ z_1^{'} & z_2^{'} & \dots & z_n^{'} \\ 1 & 1 & \dots & 1 \end{bmatrix} = \begin{bmatrix} R & t \\ 0 & 1 \end{bmatrix} \begin{bmatrix} x_1 & x_2 & \dots & x_n \\ y_1 & y_2 & \dots & y_n \\ z_1 & z_2 & \dots & z_n \\ 1 & 1 & \dots & 1 \end{bmatrix} \tag{2}$$

where the lower elements in the transformation matrix (0 and 1) are of dimensions 1×3 and 1×1 respectively.

In `mBuild`, this equivalence transform is used to force four points of one compound to overlap with four points of another. Achieving this generally, requires that the same arrangement of four non-coplanar points must be added to any compound intended to make use of the equivalence transform.

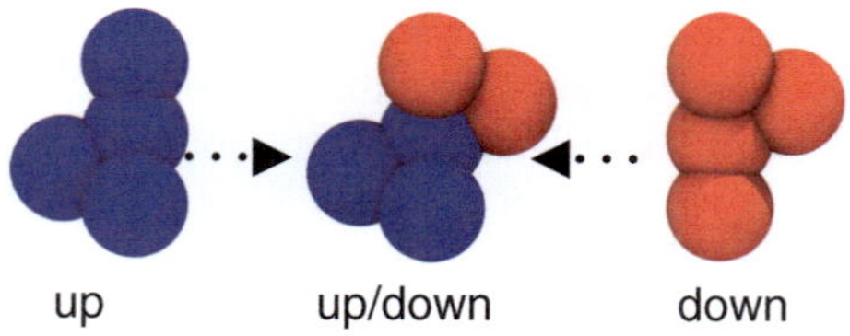

Fig. 1 The spatial arrangement of the particles within a port. Both up and down contain the same arrangement of four non-coplanar particles except that they face opposite directions

2.2.1 Ports

To formalize, simplify, and enable this behavior to function with any compound, `mBuild` provides the `Port` class, which is a simple `Compound` containing four untyped `Particles` in a compact, non-coplanar arrangement (see Fig. 1). Note that for most use cases, it is not desirable to print these untyped, extra `Particles` when outputting the final structure to a file, which is the default behavior of the `Compound.save()` method, but they can be saved if desired, e.g. for visualization purposes.

Instead of having to explicitly define an equivalence relation between four pairs of points, `mBuild` allows for declaring two `Ports`, one in each compound, to be equivalent. When performing an equivalence transform on two `Ports`, one of the `Compounds` that the two `Ports` are a part of is rotated and translated, such that the untyped particles inside their respective ports overlap (see Fig. 2). Since it is common that `Ports` represent bonding sites where molecule fragments need to be attached, `mBuild` allows for defining an *anchor* `Compound` associated with a `Port`. After the affine transformation is applied, `mBuild` will by default create a bond between the two respective anchors, relieving the user from this often tedious task.

Notice that ports have directionality, as well. Consider *Component* C_1 in Fig. 2, representing a methyl group. It is not possible to create an ethane molecule from

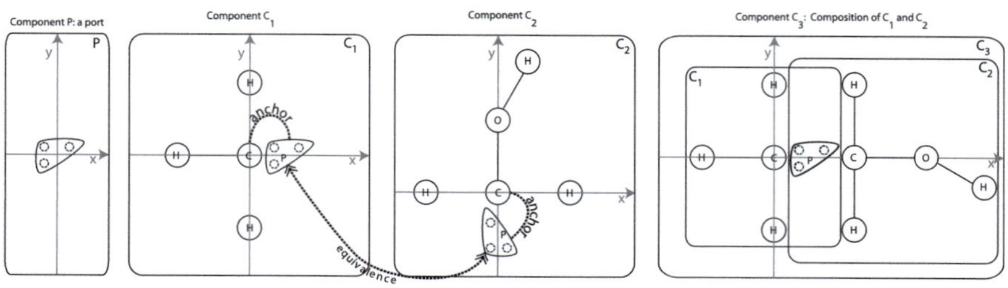

Fig. 2 A `Port` is a compound with two pairs of four `Particles`. Here, one pair of three points is shown to illustrate this 2D example. `Ports` are attached to any other `Compound`, most commonly anchored to a `Particle` where a chemical bond should exist. Compound C1 is a methyl group with a `Port` anchored to the carbon atom. C2 is a methylene bridge already connected to a hydroxyl group. C1 and C2 are then attached using the equivalence relation described in Eq. (2) to create C3, an ethanol molecule. By default, a `Bond` is created between the two anchoring carbons. Adapted with permission from Fig. 2 in Sallai, J. et al. (2013) *Web- and Cloud-based Software Infrastructure for Materials Design*. Procedia Computer Science: Elsevier

two such components, because the equivalence transform would render not just the untyped atoms in the ports, but also the carbon and hydrogen atoms to overlap. While one way of solving this problem would be to have two flavors of each such `Compound` class, one with an "outward pointing" Port, and another one with an "inward pointing" one, `mBuild` takes an alternate approach. The actual implementation of the `Port` class contains not four, but eight untyped atoms: four of them forming an "inward pointing", while the other four comprising an "outward pointing" collection of points. When performing an equivalence transform, `mBuild` computes two affine transformation matrices, and choses the one that avoids the overlap of the compounds' typed atoms. This is achieved by checking which of the two transformations forces the anchor atoms as far away from one another as possible (see Fig. 1 for an illustration of how these quartets of `Particles` are arranged). Figure 2 highlights this procedure via the construction of an ethanol molecule. Additional documentation is included at the development website (http://imodels.github.io/mbuild/) via an interactive IPython notebook [14].

3 Applications

Below, we highlight the basics of assembling low level components into successively more complex structures in `mBuild` and how to programmatically control these workflows to perform automated screening for monolayer systems. All the examples discussed below are also available as tutorials in IPython notebook format where users can seamlessly visualize components as they are constructed from Python code via a widget provided by the `imolecule` package [8]. Static versions of these notebooks are also hosted on our documentation page at http://imodels.github.io/mbuild/. Many additional example systems of varying complexity are provided together with the `mBuild` source code on GitHub.

3.1 Defining and Connecting Basic Components

The simplest way to define a basic component in `mBuild` is to draw the component using software such as Avogadro [11], output it as a `.mol2` or `.pdb` file with defined bonds and then use the `load` function in `mBuild`. Adding a Port to a compound that a user wants to be able to connect to other compounds requires placing the `Port` where a bond could be formed and specifying an anchor particle with which the `Port` is associated. Just as with any other `Compound`, `Ports` can not only be translated but also rotated thus allowing non-linear arrangements to be constructed. This procedure is highlighted in Listing 1; basic components can be stored and reused for future system construction thus minimizing the need for users to place `Ports`, as will be demonstrated as part of the construction of alkane monolayers in the screening application below.

Listing 1 Example code to generate a CH_2 group and attach two ports

```

ch2 = mb.load('ch2.pdb')
mb.translate(ch2, -ch2[0].pos) # Move carbon to origin.

port1 = mb.Port(anchor=ch2[0]) # Anchor the port on the carbon.
mb.translate(port1, [0, 0.07, 0]) # Approx. half a C-C bond length in nm.

port2 = mb.Port(anchor=ch2[0])
mb.translate(port1, [0, -0.07, 0]) # Placed on opposite side of carbon.

ch2.add(port1, label='up')
ch2.add(port2, label='down')
```

Any two `Ports` can be forced to overlap using the equivalence transform. Listing 2 demonstrates how this functionality can be leveraged via the simple yet common use case of creating an alkane polymer chain which will be used for screening—in this example, a CH_2 group with the ports "up" and "down" defined.

Listing 2 Example code for polymerizing CH_2 groups

```
import mbuild as mb
from mbuild.lib.moieties import CH2

polymer = mb.Compound()
last_monomer = CH2()

polymer.add(last_monomer)
for _ in range(10):
    this_monomer = mb.clone(last_monomer)
    mb.equivalence_transform(this_monomer, this_monomer['up'],
        last_monomer['down'])
    polymer.add(this_monomer)
    last_monomer = this_monomer
```

To further simplify the composition of basic components into more complex structures, several classes and functions have been developed to more naturally express many commonly performed tasks. For example, the functionality of the example in Listing 2 is encapsulated within the `Polymer` class which reduces the above for loop to one line for end users. For example, the PEG chains referenced in the following examples, are created with the code in Listing 3.

Listing 3 Using the Polymer class to create PEG chains

```
import mbuild as mb
from mbuild.lib.moieties import CCO

peg = mb.Polymer(CCO(), n=10)
```

3.2 Patterning Surfaces

`mBuild` provides functionality for patterning of surfaces in arbitrary ways. Below, we highlight this feature via the patterning of scientifically relevant 2D and 3D systems.

In the example shown in Fig. 3, the `TiledCompound` class is used to replicate a periodic substrate in the *x*- and *y*-dimensions. This class also internally adjusts periodic bonds. In the final tier of the hierarchy, the patterning functionality, which can be used to create patterns on, for example, substrates or spherical particles, is used to randomly disperse polymer brushes on the substrate. Functionality is provided in `mBuild` for a variety of 2D and 3D patterns including random, grid-like, disks and spherical patterns. Ultimately, a multi-tiered hierarchy of components is assembled, from simple "hand-drawn" monomers, through polymerization and replication of periodic substrates. This functionality is expressed with minimal code via creating a new Python class (shown at the bottom of Fig. 3) to expose the desirable tunable parameters. Here, the number of monomers in the chain, the number of chains on the surface, the pattern on the surface, and the size of the surface can all be trivially modified during screening.

The surface patterning illustrated in Fig. 3 was limited to a two-dimensional surface; however, the underlying functionality in `mBuild` naturally generalizes to three dimensions as well with essentially no changes to the user-level code.

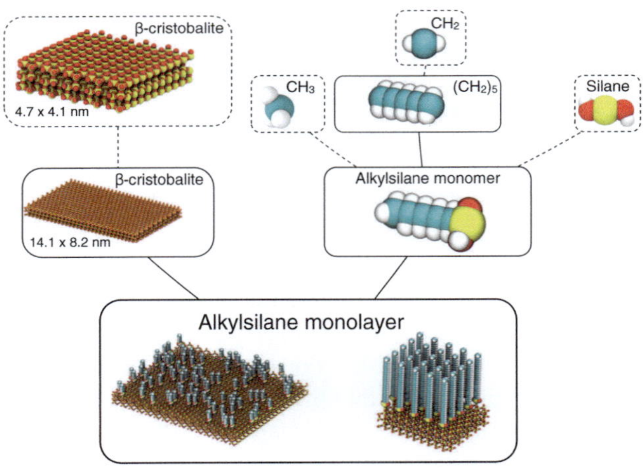


```
pattern = Random2DPattern(100)      pattern = Grid2DPattern(5, 5)
chain_length = 5                    chain_length = 20
n_tiles = [2, 3, 1]                 n_tiles = [1, 1, 1]

monolayer = AlkaneMonolayer(chain_length, pattern, n_tiles)
```

Fig. 3 Hierarchy of compounds used to generate an alkylsilane monolayer on a β-cristobalite substrate. *Dashed boxes* indicate base components for which .mol2 or .pdb files exist, e.g. drawn using software such as Avogadro [11]. The code snippet used to generate the structures with all of the tunable parameters exposed is shown at the *bottom* for two different parameter combinations

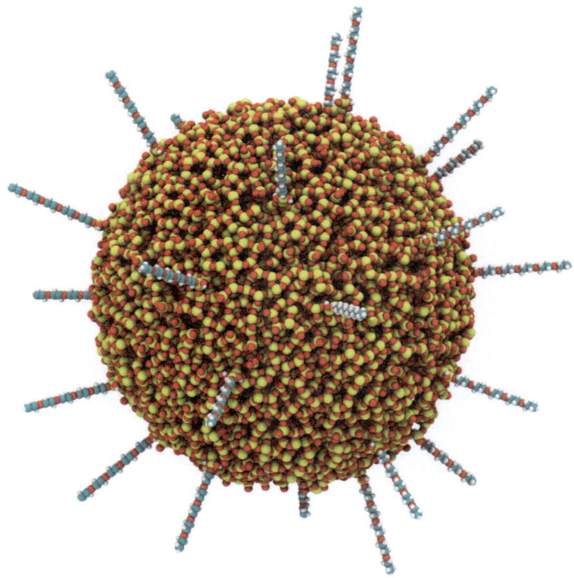

Fig. 4 An 8 nm diameter silica nanoparticle sparsely functionalized with PEG chains bound to the surface with a silane group

Figure 4 and Listing 4 show how this could be used to functionalize a spherical nanoparticle with various polymer chains. The code utilized to attach chemical groups to two-dimensional systems can be reused for three dimensional structures without significant modification or further effort by the end user.

Listing 4 Example code to tether PEG chains to a silica nanoparticle

```
import mbuild as mb
from mbuild.lib.surfaces import SilicaNP
from mbuild.examples import PEGSilane

peg_silane = PEGSilane(peo_units=5) # Length based on number of CCO moieties.
silica_np = SilicaNP()

pattern = mb.SpherePattern(25)
peg_chains, _ = pattern.apply_to_compound(guest=peg_silane, host=silica_np)

tnp = mb.Compound([silica_np, peg_chains])
```

4 Screening Soft Matter Systems: Self-assembled Monolayers

Building upon the prior examples, monolayers are constructed in a programmatic way to demonstrate the use of `mBuild` for screening applications. Monolayers encompass a vast chemical parameter space that can be tuned for applications such as lubrication [15] and anti-fouling [16], and their behavior and properties often strongly depend on the substrate, binding moiety, chain type, composition of multiple chain types, surface patterning, etc. Sampling more than one or two dimensions of this parameter space using experimental techniques, while technically possible, quickly becomes limited by practical considerations. Molecular dynamics can be used as a screening step to inform subsequent experimental studies and dramatically cut down the relevant search space. Here, using `mBuild` substrate density, chain length, and chain type of monolayer systems are programmatically varied in order to perform a basic screening.

The first step to performing a screening procedure across chemical space involves building the input topologies. Ideally, a user should have seamless access to any variables of interest thus enabling them to adjust these to mimic a statistical distribution. As discussed previously, the hierarchical nature of `mBuild` provides an avenue to expose an arbitrary set of variables to the end user and thus enables users to leverage the scientific Python ecosystem to apply standard optimization techniques and analysis to explore chemical parameter space. As highlighted above, in `mBuild`, the only explicit rotation and translation occurs in the lowest level of the hierarchy when placing ports. Once these simple components have been fitted with ports, they can be stored in the database for future use thus completely eliminating the need for explicit rotation and translation when building many systems; here, we reuse many of the components previously defined in the prior examples. Each higher tier in the hierarchy contains only a few lines of code to express which ports to connect to one another.

Listing 5 shows the `mBuild` code that generates configurations for a simple screening procedure of alkane and PEG monolayers on silica substrates. This code varies the chain length and the number of chains on the surface for both molecules types. It is important to note that the code to generate both monolayer types are nearly identical due to the hierarchical nature of `mBuild`; the `Monolayer` function is generic, as it simply expects a `Compound` with a `Port` defined for attachment. Thus it can readily accept either the `Alkane` or `PEG Compounds` (or mixture thereof) that have previously been define, where each of these `Compounds` accepts an argument to define the length of the desired polymer chain. As such, this example can be trivially extended by creating a different molecule `Compound`, and substituting this in place of either the `Alkane` or `PEG Compound`.

Figure 5 illustrates two of the systems created using this procedure post-equilibration. In this example, the monolayers were patterned in a 2D grid but the patterning of the surface is also tunable if desired, as shown previously. Each monolayer that was created was sampled for 10 ns using GROMACS [17] and the OPLS-aa forcefield [18] with modifications as described by Lorentz et al. [19].

Listing 5 Example code to generate alkane and PEG monolayers differing in both chain length and number of surface grafted chains. Note that most of the code can be reused to create both the PEG and alkane monolayer; the only difference is the chain class that is instantiated

```python
import mbuild as mb
from mbuild.examples import Alkane, PEG
from mbuild.lib.atoms import H
from mbuild.lib.surfaces import Betacristobalite

hydrogen = H()
surface = Betacristobalite()

for sqrt_n_chains in range(4, 11): # Amount of chains on surface.
    pattern = mb.Grid2DPattern(sqrt_n_chains, sqrt_n_chains)

    for chain_length in range(6, 22, 3): # Length of chains on surface.
        chain = Alkane(chain_length) # Use alkane chains.
        monolayer = mb.Monolayer(surface, chain, pattern, backfill=hydrogen)
        monolayer.save('{}_{}_alkane.mol2'.format(sqrt_n_chains**2, cl))

        chain = PEG(peo_units=chain_length / 3) # Use PEG chains.
        monolayer = mb.Monolayer(surface, chain, pattern, backfill=hydrogen)
        monolayer.save('{}_{}_peg.mol2'.format(sqrt_n_chains**2, cl))
```

Figure 6 shows the average nematic order parameter, *S*2, of the chains on monolayer [20, 21]. *S*2 measures the orientational ordering of the chains, where for monolayers, values below 0.7 indicate a fluid-like state (i.e., low order) whereas values that approach unity indicate a high degree of crystalline orientational ordering. It has been shown that *S*2 influences the frictional properties of monolayers, where lower values of *S*2 for monolayers tend to be correlated with higher frictional forces when the monolayers are brought together in sliding contact [22]. Thus *S*2 serves as a useful surrogate for rapidly screening monolayers to determine which regimes are likely to produce high/low coefficients of frictions. While additional simulations and sampling are required to draw more robust conclusions, several regimes are readily apparent. A clear transition from disordered, fluid-like monolayer states to ordered states occurs for both systems. This transition occurs at lower surface coverages for alkane chains as compared to PEG chains. That is, PEG systems appear to have a smaller regime of well order states, which can be

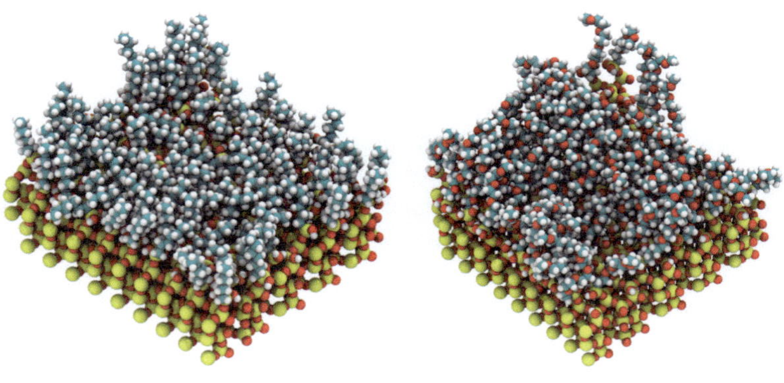

Fig. 5 An alkane system with 81 chains with 7 carbons each (*left*) and a PEG system with 64 chains and 13 carbons/oxygens (*right*). Both shown post-equilibration

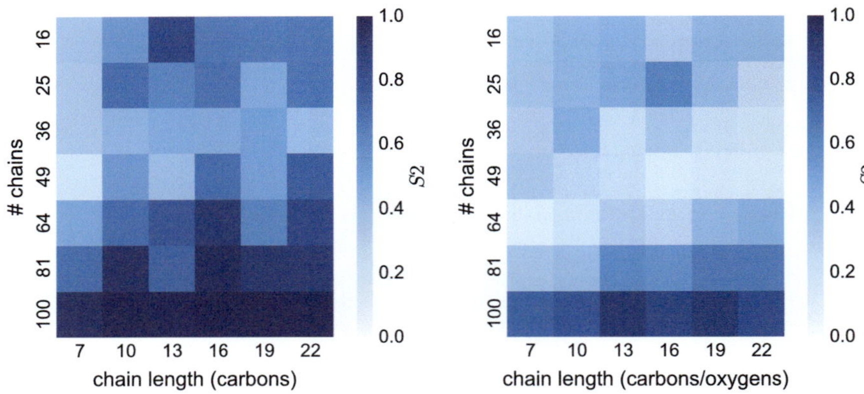

Fig. 6 Average nematic order parameter of every system after 10 ns of sampling. The total process of constructing all 84 systems with `mBuild` takes a few minutes on a modern laptop and the simulations each take approximately 0.5–3 h depending on system size using a GTX980 and 8 CPU cores of an Intel Xeon E5 2600v3

accounted for by the increased flexibility in PEG. In both cases, systems with the highest values of *S*2 tend to occur for higher surface densities and longer chains, and thus one would expect materials in these regimes to demonstrate the most favorable frictional properties. Interestingly, these screening simulations also reveal a second regime for PEG occurring for low surface coverage and short chain length; in this regime moderate values of *S*2 are observed, which, upon visual inspection, appears associated with chains lying flat along the surface. The ability to rapidly screen, evaluate and cross-correlate metrics like the nematic order parameter will accelerate our ability to rationally design soft materials in complex parameter landscapes.

5 Conclusion

`mBuild` provides a programmatic pathway to constructing arbitrary, complex input topologies for molecular simulations. The use of an equivalence operator typically eliminates the need for users to explicitly rotate or translate components while assembling chemical structures. The core data structures of `mBuild` and how the equivalence operator is implemented and used in practice are described and the pathway from basic component creation all the way through constructing several complex example hierarchies illustrated. The format-agnostic nature of `mBuild` allows for flexible interoperability with other tools in the scientific Python and molecular modeling communities, such as packmol [23], polymatic [6], MDTraj [7], imolecule [8], OpenMM [9] and HOOMD-blue [10]. Using monolayers as an example, the power of this approach is highlighted by performing a small parameter sweeping simulation study, demonstrating clear regimes of highly ordered monolayers which are likely correlated with favorable friction coefficients. This example demonstrates how this approach can be leveraged to more broadly study, design and optimize complex materials. Source code and interactive tutorials in the IPython notebook format, which reinforce the basics of component construction and how to re-use components to assemble more complex systems, are also provided on the `mBuild` website (http://imodels.github.io/mbuild/).

The amount of easily generatable chemical configurations scales dramatically as users contribute components to `mBuild`'s library. As such, we have begun curating a version-controlled library of components such that they can be reused, error-corrected and added to. `mBuild` and its component library are fully open-sourced at https://github.com/imodels/mbuild and user contributions are actively encouraged, which we hope will attract an active user base.

Acknowledgments This material is based upon work supported by the National Science Foundation under Grants No. NSF CBET-1028374 and OCI-1047828.

References

1. Bernstein, F.C., Koetzle, T.F., Williams, G.J., Meyer, E.F., Brice, M.D., Rodgers, J.R., Kennard, O., Shimanouchi, T., Tasumi, M.: The protein data bank: a computer-based archival file for macromolecular structures. Arch. Biochem. Biophys. **185**, 584–591 (1978)
2. Humphrey, W., Dalke, A., Schulten, K.: VMD: visual molecular dynamics. J. Mol. Graph. **14**, 33–38 (1996)
3. Salomon-Ferrer, R., Case, D.A., Walker, R.C.: An overview of the Amber biomolecular simulation package. Wiley Interd. Rev.: Comput. Mol. Sci. **3**, 198–210 (2013)
4. Omnia: High performance, high usability toolkits for predictive biomolecular simulation. http://www.omnia.md
5. Sallai, J., Varga, G., Toth, S., Iacovella, C.T., Klein, C., McCabe, C., Ledeczi, A., Cummings, P.T.: Web- and cloud-based software infrastructure for materials design. Proc. Comput. Sci. **29**, 2034–2044 (2014)
6. Abbott, L.J., Hart, K.E., Colina, C.M.: Polymatic: a generalized simulated polymerization algorithm for amorphous polymers. Theoret. Chem. Acc. **132**, 1–19 (2013)
7. McGibbon, R.T., Beauchamp, K.A., Schwantes, C.R., Wang, L.-P., Hernández, C.X., Harrigan, M.P., Lane, T.J., Swails, J.M., Pande, V.S.: MDTraj: a modern, open library for the analysis of molecular dynamics trajectories. bioRxiv (2014)
8. Fuller, P.: Imolecule: an embeddable webGL molecule viewer. https://github.com/patrickfuller/imolecule
9. Eastman, P., et al.: OpenMM 4: a reusable, extensible, hardware independent library for high performance molecular simulation. J. Chem. Theory Comput. **9**, 461–469 (2013)
10. Anderson, J.A., Lorenz, C.D., Travesset, A.: General purpose molecular dynamics simulations fully implemented on graphics processing units. J. Comput. Phys. **227**, 5342–5359 (2008)
11. Hanwell, M.D., Curtis, D.E., Lonie, D.C., Vandermeersch, T., Zurek, E., Hutchison, G.R.: Avogadro: an advanced semantic chemical editor, visualization, and analysis platform. J. Cheminform. **4**, 17 (2012)
12. http://www.whitehouse.gov/mgi. Materials genome initiative for global competitiveness

13. Aric Hagberg, P.S., Dan Schult NetworkX: High-productivity software for complex networks. https://networkx.github.io/
14. Pérez, F., Granger, B.E.: IPython: a system for interactive scientific computing. Comput. Sci. Eng. **9**, 21–29 (2007)
15. Bhushan, B., Israelachvili, J.N., Landman, U.: Nanotribology: friction, wear and lubrication at the atomic scale. Nature **374**, 607–616 (1995)
16. Brzoska, J.B., Shahidzadeh, N., Rondelez, F.: Evidence of a transition temperature for the optimum deposition of grafted monolayer coatings. Nature **360**, 719–721 (1992)
17. Pronk, S., Páll, S., Schulz, R., Larsson, P., Bjelkmar, P., Apostolov, R., Shirts, M.R., Smith, J. C., Kasson, P.M., Van Der Spoel, D., Hess, B., Lindahl, E.: GROMACS 4.5: a high-throughput and highly parallel open source molecular simulation toolkit. Bioinformatics **29**, 845–854 (2013)
18. Jorgensen, W.L., Maxwell, D.S., Tirado-Rives, J.: Development and testing of the OPLS all-atom force field on conformational energetics and properties of organic liquids. J. Am. Chem. Soc. **118**, 11225–11236 (1996)
19. Lorenz, C., Webb, E., Stevens, M., Chandross, M., Grest, G.: Frictional dynamics of perfluorinated self-assembled monolayers on amorphous SiO_2. Tribol. Lett. **19**, 93–98 (2005)
20. Lagomarsino, M.C., Dogterom, M., Dijkstra, M.: Isotropic nematic transition of long, thin, hard spherocylinders confined in a quasi-two-dimensional planar geometry. J. Phys. Chem. **119**, 719–721 (2003)
21. Wilson, M.R.: Determination of order parameters in realistic atom-based models of liquid crystal systems. J. Mol. Liq. **68**, 23–31 (1996)
22. Black, J.E., Iacovella, C.R., Cummings, P.T., McCabe, C.: Molecular dynamics study of alkylsilane monolayers on realistic amorphous silica surfaces. Langmuir **31**, 3086–3093 (2015)
23. Martnez, L., Andrade, R., Birgin, E.G., Martnez, J.M.: PACKMOL: a package for building initial configurations for molecular dynamics simulations. J. Comput. Chem. **30**, 2157–2164 (2009)

Quantum Virial Coefficients via Path Integral Monte Carlo with Semi-classical Beads

Ramachandran Subramanian, Andrew J. Schultz and David A. Kofke

Abstract Conventionally, Path Integral Monte Carlo (PIMC) calculations are performed with 'classical beads' (beads interacting via a classical potential) by using the primitive approximation for the thermal density matrix. Higher order propagators of the thermal density matrix have been proven to achieve faster convergence and better precision in quantum calculations than using just the primitive approximation. Use of different propagators in PIMC leads to methods equivalent to performing PIMC with 'semi-classical beads' (beads interacting via a semi-classical potential). We examine the Takahashi-Imada (TI) propagator as well as an ad hoc semi-classical potential in PIMC calculations for computing the quantum second virial coefficient for helium-4. We compare the performance of the two approaches based on semi-classical beads against values computed from PIMC using conventional classical beads. We find that while the TI propagator has the same or marginally better precision compared to the classical case, it has the best convergence rate (with respect to number of path-integral beads) among the three approaches. The convergence rate of the ad hoc potential is marginally better than its classical counterpart, and its precision is approximately the same as the classical case.

Keywords Path integral Monte Carlo · Takahashi-Imada propagator · Quantum virial coefficients · Helium-4 · Thermal density matrix

1 Introduction

The thermal density matrix ρ plays a key role in Feynman's imaginary-time Path Integrals (PI) formalism and its application in Monte Carlo (MC) algorithms to compute physical properties of interest. In position space, it is given by [1–3]:

R. Subramanian · A.J. Schultz · D.A. Kofke (✉)
Department of Chemical and Biological Engineering, University at Buffalo,
The State University of New York, Buffalo, NY 14260-4200, USA
e-mail: kofke@buffalo.edu

R.Q. Snurr et al. (eds.), *Foundations of Molecular Modeling and Simulation*,
Molecular Modeling and Simulation, DOI 10.1007/978-981-10-1128-3_6

$$\rho(R, R'; \beta) = \langle R | e^{-\beta \mathcal{H}} | R' \rangle \tag{1}$$

where $R = \{\boldsymbol{r}_1, \boldsymbol{r}_2, \ldots \boldsymbol{r}_n\}$ and $\beta = 1/k_B T$, with k_B Boltzmann's constant and T the temperature. A key property of the density matrix is that the product of two density matrices is also a density matrix:

$$\rho(R, R'; \beta_1) \times \rho(R, R'; \beta_2) = \rho(R, R'; \beta_1 + \beta_2) \tag{2}$$

This is because any operator (specifically the Hamiltonian operator $\mathcal{H}$ here) is commutative with any scalar multiple of itself. This exact property allows us to write down the following $(P-1)$-fold convolution:

$$\rho(R_0, R_P; \beta) = \int \cdots \int \mathrm{d}R_1\, \mathrm{d}R_2 \ldots \mathrm{d}R_{P-1}\, \rho(R_0, R_1; \tau)\rho(R_1, R_2; \tau)\ldots\rho(R_{P-1}, R_P; \tau) \tag{3}$$

where $\tau = \beta/P$. Note that even though the above expression is exact, one needs to make approximations to the thermal density matrix in order to compute the convolution efficiently. The simplest of the approximations is to assume that the kinetic-energy operator ($\mathcal{T}$) and the potential-energy operator ($\mathcal{V}$) in the Hamiltonian commute with each other. As $\tau \to 0$ or equivalently as $PT \to \infty$, the "primitive approximation" is given by:

$$e^{-\tau(\mathcal{T}+\mathcal{V})} \approx e^{-\tau\mathcal{T}} e^{-\tau\mathcal{V}} \tag{4}$$

The Trotter formula proves that this approximation does converge to the right result in the $P \to \infty$ limit and is given by:

$$e^{-\beta(\mathcal{T}+\mathcal{V})} = \lim_{P\to\infty} [e^{-\tau\mathcal{T}} e^{-\tau\mathcal{V}}]^P \tag{5}$$

It is worth noting that within the PI implementation, we are mainly interested in evaluating the trace of the density matrix, as it is directly related to the partition function. Also when using the primitive approximation, we neglect terms that are of the order τ^2. To improve the precision of results in MC simulations and to achieve faster convergence as P increases, higher order corrections (or propagators of the density matrix) have been developed.

The Takahashi-Imada (TI) propagator [4] with error of the order τ^4 uses:

$$\begin{aligned} Tr\left[e^{-\beta(\mathcal{T}+\mathcal{V})}\right] &= Tr\left[e^{-\frac{\beta}{P}\mathcal{T}} e^{-\frac{\beta}{P}\mathcal{V}'}\right]^P + O(\beta^5 P^{-4}), \\ \mathcal{V}' &= \mathcal{V} + \frac{1}{24}\left(\frac{\beta}{P}\right)^2 [\mathcal{V}, [\mathcal{T}, \mathcal{V}]]. \end{aligned} \tag{6}$$

Given a system with Hamiltonian $\mathcal{H}$ as:

$$\begin{aligned}\mathcal{H} &= \mathcal{T} + \mathcal{V},\\ \mathcal{T} &= -\frac{\hbar^2}{2m}\sum_{i=1}^{N}\frac{\partial^2}{\partial r_i^2},\\ \mathcal{V} &= V(\boldsymbol{r}_1,\ldots,\boldsymbol{r}_N),\end{aligned} \tag{7}$$

it can be easily shown that from Eqs. (6) and (7), we get the following:

$$\mathcal{V}' = V(\boldsymbol{r}_1,\ldots,\boldsymbol{r}_N) + \frac{\hbar^2}{24m}\left(\frac{\beta}{P}\right)^2\sum_{i=1}^{N}|\nabla_i V(\boldsymbol{r}_1,\ldots,\boldsymbol{r}_N)|^2, \tag{8}$$

where $\hbar$ is the reduced Planck's constant and ∇_i denotes the gradient with respect to coordinates of the ith atom. Equations (6) and (8) constitute the working equations of the TI propagator. Schenter [5] computed fully quantum virial coefficients using three different interaction potentials for water and found that using the semi-classical TI approximation (Eq. (8) with $P = 1$) gave the best agreement to fully quantum statistical mechanical calculations, especially at low temperatures where conventional expressions (based on the primitive approximation) including first order quantum corrections failed.

Janke and Sauer [6] showed that by adopting a slightly modified version of the Trotter formula (Eq. 5), they could systematically decrease the variance of the propagator. By decomposing the Hamiltonian to include more and more components of the kinetic- and potential-energy operators, they observed that the variance of the propagator improved. Suzuki [7] suggested new schemes for the exponential product formulae along with a basic theorem for a generalized decomposition that results in the propagator having error of the order $O(1/P^4)$. Yamamoto [8] showed that using a finite-difference based approach (instead of computing derivatives involved with the use of TI and Suzuki propagators) helped improve the variance further.

In this paper, we compute fully quantum virial coefficients of helium-4 using the TI propagator (Eqs. 6–8); we also consider the use of an ad hoc semi-classical potential (details of which will be explained in Sect. 4). Calculation of very precise physical properties of helium is of interest in the field of metrology to develop accurate calibration and pressure standards, to accurately compute the Boltzmann constant, and to improve acoustic gas thermometry [9–14]. Semi-classical virial coefficients up to fifth order have been computed for helium-4 by Shaul et al. [15], and showed that first-principles properties could be evaluated with precision and accuracy that exceeds experiment. Garberoglio and Harvey [9, 16, 17] reported fully quantum second and third virial coefficients for helium-3 and helium-4 including exchange effects where needed, for temperatures as low as 2.6 K. Shaul et al. [18] reported fully quantum virial coefficients of helium-4 (but without exchange) up to fourth order for temperatures of T = 2.6–1000 K.

The present application is primarily interested in demonstrating a variation of the PIMC methodology, rather than establishing new or more precise values of virial coefficients of helium. In Sect. 2 we will introduce the basics of computing quantum virial coefficients using PIMC. Sect. 3 explains how the thermal density matrix is used in PIMC to compute quantum virial coefficients. Sect. 4 contains all the simulation details including the ab initio Potential Energy Surface (PES) used, the range of temperatures investigated and other relevant computational parameters. Sect. 5 discusses the results and its comparison with values from literature, and also examines the performance of the various approaches used. We provide concluding remarks and ideas for future work in Sect. 6.

2 Quantum Virial Coefficients

Virial coefficients are important thermodynamic quantities of a system for two main reasons:

- They lead to other physical properties such as the pressure, critical temperature etc.
- They can be evaluated computationally given an interaction potential, and also by experiments. Thus the accuracy of the interaction potential can be judged based on the accuracy of the virial coefficients relative to experimental results.

The virial coefficients are generally denoted as B_N and the first two virial coefficients are given by [19]:

$$\begin{aligned} B_2(T) &= -\frac{1}{2!\underline{V}}\left[Z_2^* - Z_1^{*2}\right], \\ B_3(T) &= -\frac{1}{3!\underline{V}^2}\left[\underline{V}(Z_3^* - 3Z_2^*Z_1^* + 2Z_1^{*3}) - 3(Z_2^* - Z_1^{*2})^2\right], \end{aligned} \tag{9}$$

where $Z_N^* \equiv \left[N!\left(\frac{V}{Q_1}\right)^N Q_N\right]$ is the N-body configurational integral, Q_N is the N-body canonical partition function, and $\underline{V}$ is the volume.

The N-body configurational integral, which depends on the N-body interaction potential, becomes exponentially more difficult to compute with increasing N. For extremely simple interaction potentials like hard spheres, up to fourth-order virial coefficients may be calculated analytically [20]. Higher order virial coefficients using more complicated interaction potentials need to be evaluated numerically through quadrature or by using MC simulations. Upon further simplification and assuming pairwise additivity of the potential, we can rewrite Eq. (9) using Mayer f-functions as [20, 21]:

$$\begin{aligned} B_2(T) &= -\frac{1}{2}\int \mathrm{d}1\, f(0,1), \\ B_3(T) &= -\frac{1}{3}\iint \mathrm{d}1\ \mathrm{d}2\, f(0,1)\, f(0,2)\, f(1,2), \end{aligned} \tag{10}$$

where $f(0,1) = (\exp[-\beta U_2(\boldsymbol{r})] - 1)$ and indices '1' and '2' denote the position and orientational degrees of freedom of molecules 1 and 2, respectively, with respect to molecule '0' at the origin.

Empirical potentials, which are usually functions that are fit to experimental data, tend to predict the net effect of a variety of phenomena over a range of conditions, and are consequently less accurate than ab initio PES for describing *N*-body interactions. The virial coefficients that are calculated from an input interaction potential (empirical or ab initio PES) without modification are known as classical virial coefficients because they do not include nuclear quantum effects explicitly. Virial coefficients computed using an effective potential such as the Quadratic Feynman-Hibbs (QFH) [1] that includes a quantum correction are known as semi-classical virial coefficients.

Nuclear quantum effects are almost always ignored in the development of an ab initio PES because of the Born-Oppenheimer approximation, which greatly simplifies the electronic Schrödinger equation by separating or decoupling the coordinates of the electrons from those of the nuclei. However, quantum mechanics prescribes that the wave functions of the atoms/molecules become more diffuse at low temperatures, or in other words, they become "fuzzy." This behavior has an effect on the virial coefficient. Therefore, when using ab initio PESs for the calculation of physical properties, especially at low temperatures, one needs to account for the nuclear quantum effects explicitly. The discretized PI formalism of Feynman provides a route to approximate the inherent fuzziness of an atom/molecule at low temperatures as a closed ring of 'beads' that represent the atom/molecule at *P* different imaginary-time instances. The formalism maps the quantum mechanical partition function onto the classical partition function of a closed ring polymer with *P* beads where adjacent beads are connected by harmonic springs whose stiffness depends on the temperature, atomic mass and *P*. The larger the discretization parameter *P*, the better the characterization of the fuzziness.

PIMC involves simulating different configurations of the closed ring polymer and accepting/rejecting it based on some MC criteria. The property of interest (usually the interaction potential) is then averaged across the simulation with each configuration having an appropriate weight. The interaction potential between two molecules is defined to be the average of the inter-molecular potential energy over corresponding beads of the two rings. Virial coefficients that are calculated from an input interaction potential including nuclear quantum effects using PIMC method are therefore known as (fully) quantum virial coefficients.

3 Thermal Density Matrix and PIMC

In this section, we will show how the thermal density matrix is used in PIMC to compute quantum virial coefficients. Consider the Hamiltonian of a monatomic molecule like helium with mass m (Eq. 7). Using the primitive approximation (Eq. 4), Trotter formula (Eq. 5), and following the procedure outlined in Ref. [9], we can obtain the kinetic-energy operator matrix elements as:

$$\left\langle \boldsymbol{r}_i \middle| \exp\left(-\frac{\beta \hat{p}^2}{2mP}\right) \middle| \boldsymbol{r}_j \right\rangle = \frac{P^{3/2}}{\Lambda^3} \exp\left(-\frac{K(\boldsymbol{r}_i - \boldsymbol{r}_j)^2}{2}\right), \tag{11}$$

where $K = \frac{2\pi P}{\Lambda^2}$, $\Lambda = \frac{h}{\sqrt{2\pi m k_B T}}$.

The potential-energy operator matrix elements can similarly be written as [3]:

$$\left\langle \boldsymbol{r}_i \middle| \exp\left(-\frac{\beta V(\boldsymbol{r})}{P}\right) \middle| \boldsymbol{r}_j \right\rangle = \exp\left(-\frac{\beta V(\boldsymbol{r}_i)}{P}\right) \delta(\boldsymbol{r}_i - \boldsymbol{r}_j). \tag{12}$$

It can be easily shown [9] then that the expression for the fully quantum second virial coefficient can be written as:

$$B_2(T) = -2\pi \int \mathrm{d}r\, r^2 (e^{-\beta V_{2,\mathrm{eff}}(r)} - 1), \tag{13}$$

where

$$e^{-\beta V_{2,\mathrm{eff}}(r)} = \int \prod_{i=1}^{P-1} \mathrm{d}^3 \Delta \boldsymbol{r}_i\, e^{-\beta \overline{U}_2(r)}\, F_{\mathrm{ring}}(m; \Delta \boldsymbol{r}_1, \ldots, \Delta \boldsymbol{r}_{P-1}), \tag{14}$$

$$\begin{gathered} \overline{U}_2(r) = \frac{1}{P} \sum_{i=1}^{P} U_2(\boldsymbol{r}_{1,i}, \boldsymbol{r}_{2,i}), \\ |r|^2 = |\boldsymbol{r}_1^{cm} - \boldsymbol{r}_2^{cm}|^2,\ \boldsymbol{r}_i^{cm} \equiv \frac{1}{P} \sum_{j=1}^{P} \boldsymbol{r}_{i,j} \end{gathered} \tag{15}$$

$$\begin{gathered} F_{\mathrm{ring}}(m; \Delta \boldsymbol{r}_1, \ldots, \Delta \boldsymbol{r}_{P-1}) = \Lambda^3 \left(\frac{P^{3/2}}{\Lambda^3}\right)^P \exp\left[-\frac{K}{2} \sum_{i=1}^{P} \Delta \boldsymbol{r}_i^2\right], \\ \Delta \boldsymbol{r}_i = \boldsymbol{r}_{i+1} - \boldsymbol{r}_i\ (i = 1, \ldots, P-1). \end{gathered} \tag{16}$$

Here $F_{\mathrm{ring}}(m; \Delta \boldsymbol{r}_1, \ldots, \Delta \boldsymbol{r}_{P-1})$ represents the weight of a ring polymer configuration, $U_2(\boldsymbol{r}_{1,i}, \boldsymbol{r}_{2,i})$ is the inter-molecular potential energy between the ith beads of rings 1 and 2, $V_{2,\mathrm{eff}}(r)$ is an effective inter-molecular potential defined by Eq. (14) and r is the inter-molecular separation.

The kinetic-energy operator in the Hamiltonian gives rise to the weight of the ring configuration, which depends on the harmonic energy of the system with P beads. The potential-energy operator (and hence, the ab initio PES) leads to the effective potential $V_{2,\text{eff}}(r)$ in the expression for the quantum virial coefficient. Recall that we used the primitive approximation where the potential-energy operator was a simple function of the PES. If instead, we were to include higher order terms in the primitive approximation using the TI propagator, we would expect it to affect only $\overline{U}_2(r)$. This would in turn lead to a change in the effective potential $V_{2,\text{eff}}(r)$. From Eqs. (7) and (8), Eq. (15) can be rewritten as follows:

$$\overline{U}_2(r) = \frac{1}{P}\sum_{i=1}^{P}\left[U_2(\boldsymbol{r}_{1,i},\boldsymbol{r}_{2,i}) + \frac{\hbar^2}{24m}\left(\frac{\beta}{P}\right)^2 |\nabla U_2(\boldsymbol{r}_{1,i},\boldsymbol{r}_{2,i})|^2\right]. \tag{17}$$

We can see that the argument within the sum on the right-hand side of the expression for $\overline{U}_2(r)$ goes from being a quantity completely independent of P and $\hbar$ as in Eq. (15) to a quantity that is dependent on both P and $\hbar$ as in Eq. (17). The inter-molecular potential experienced by the beads of the ring changes from being classical to semi-classical (dependent on P and $\hbar$). Therefore, the phrase 'PIMC with **semi-classical beads**' along with Fig. 1 is an apt description of such a PIMC

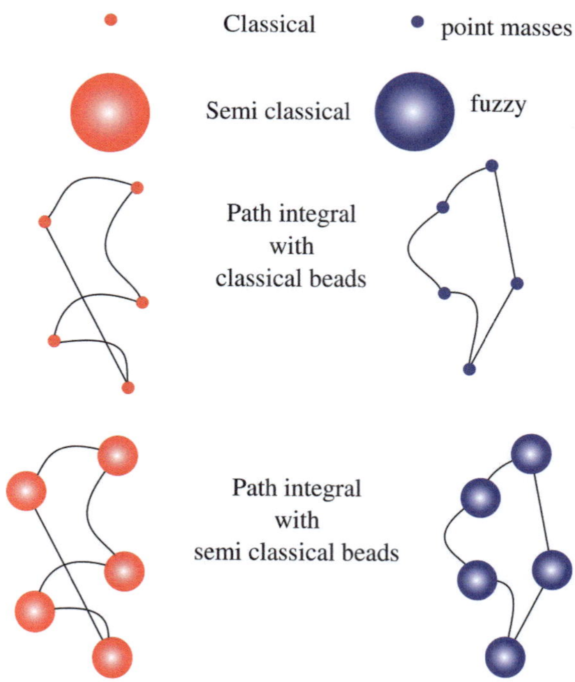

Fig. 1 Different levels of "quantumness" of a B_2 calculation going from classical virial coefficients that are calculated assuming point masses to fully quantum virial coefficients with semi-classical beads. The different sphere sizes here are for illustrative purposes only and no quantitative inference should be made

simulation. For a fixed P, we would expect to capture more quantum effects with the use of semi-classical beads than classical beads, that is, by using the primitive approximation with higher order terms than just the primitive approximation by itself.

4 Computational Details

In Sect. 3 we noted that using different propagators brought about changes only in the effective potential used. While a given propagator will correspond to some effective potential, the converse might not necessarily be true—selection of an ad hoc effective potential might not map back to an appropriate propagator. Still, it may be interesting to examine other choices of semi-classical potential for use in a PIMC framework, without deriving it from a propagator. Once the accuracy of such an ad hoc potential is established empirically, we can then compare its efficiency against the TI propagator. We have in mind in particular the QFH effective potential [1, 5], modified slightly for this purpose. We denote this as QFH* and it is given as:

$$U_2^{\mathrm{QFH}^*}(\boldsymbol{r}_{1,i},\boldsymbol{r}_{2,i}) = U_2(\boldsymbol{r}_{1,i},\boldsymbol{r}_{2,i}) + \frac{\hbar^2\beta}{24mP^2}\left[\frac{\partial^2 U_2(\boldsymbol{r}_{1,i},\boldsymbol{r}_{2,i})}{\partial r_{12,i}^2} + \frac{2}{r_{12,i}}\frac{\partial U_2(\boldsymbol{r}_{1,i},\boldsymbol{r}_{2,i})}{\partial r_{12,i}}\right],$$
$$|r_{12,i}|^2 = |\boldsymbol{r}_{1,i} - \boldsymbol{r}_{2,i}|^2 \tag{18}$$

where m is the mass of the atom. We use the $1/P^2$ prefactor for the second term as it closely resembles the TI propagator and also gives the best results of those we examined. The standard QFH semi-classical potential is obtained for $P = 1$.

The ab initio helium pair potential that we used is due to Przybytek et al. [22] (denoted as u) and a simplified, approximate version of the same (denoted as u^{simple}) was obtained from supplementary material of Shaul et al. [18]. We investigated a total of 8 temperatures ranging from T = 2.5–500 K. Mayer Sampling Monte Carlo (MSMC) [23, 24], which uses importance sampling to compute virial coefficients efficiently for any given interaction potential, was employed in our calculations.

Since this work is aimed at extending the work of Shaul et al. [18], we shall be comparing the performance of the TI propagator and the QFH* effective potential against their results. In order to make a fair and consistent comparison, we employ the same decomposition algorithms as Shaul et al. [18]. These schemes were developed to improve the efficiency of the virial coefficient calculation, doing so by computing the full quantum virial coefficients through a series of stages of increasing accuracy in the quantum treatment and adherence to the target PES. We have the same three choices for the preliminary approximation: (1) semi-classical, $[\Gamma^{\mathrm{SCL}}(u)]$, (2) the u^{simple} approximation to the semi-classical treatment, $[\Gamma^{\mathrm{SCL}}(u^{\mathrm{simple}})]$ and (3) the u^{simple} approximation to u for a finite P, $[\Gamma(P,u^{\mathrm{simple}})]$.

Here Γ represents the configurational integral of the associated potential, the square brackets indicate an independent simulation, and the superscript SCL denotes a QFH semi-classical approximation (Eq. 18) to u or u^{simple}. Since we are interested only in B_2, the Percus-Yevick compressibility route approximation [21, 25, 26] to the semi-classical approximation is ignored.

Any computational details regarding the inter-molecular potentials, decomposition strategies and MSMC parameters that are not included here may be found in Sect. B of Shaul et al. [18] and the supplementary material therein.

5 Results

For ease of reference and use, we note and define the following:

- All the simulations involved the same set of inter-molecular potentials, u or u^{simple} or their semi-classical approximations.
- We denote quantum virial coefficient results from Shaul et al. [18] as B_2^{cl}, those using the QFH* effective potential as $B_2^{\mathrm{sc,QFH^*}}$, and those using the TI propagator as $B_2^{\mathrm{sc,TI}}$; the first part of the superscript denotes which type of beads (classical (cl) or semi-classical (sc)) were used in the PIMC calculations. Note this is not to be confused with the preliminary approximations $[\Gamma^{\mathrm{SCL}}(u)]$ or $[\Gamma^{\mathrm{SCL}}(u^{\mathrm{simple}})]$ which denote the semi-classical calculations using the QFH approximations to u and u^{simple} respectively.
- In the same spirit, we refer to the algorithm for computing B_2^{cl} as the Classical Beads approach denoted as CB; $B_2^{\mathrm{sc,QFH^*}}$ as the Semi-Classical Beads QFH* approach denoted as SCB-QFH*; $B_2^{\mathrm{sc,TI}}$ as the Semi-Classical Beads TI approach denoted as SCB-TI.
- It is possible to use a semi-classical TI approximation ($P = 1$ in Eq. (8)) instead of the QFH (Eq. 18) approximation while using the TI propagator. However, after performing several calculations, we observed that using semi-classical TI approximations as preliminary approximations always led to inefficient decompositions, which resulted in larger uncertainties in $B_2^{\mathrm{sc,TI}}$ than B_2^{cl} or $B_2^{\mathrm{sc,QFH^*}}$. This is because the uncertainty of the quantity $y = [\Gamma(P, u^{\mathrm{simple}}) - \Gamma^{\mathrm{SCL}}(u^{\mathrm{simple}})]$, was significantly higher when using the semi-classical TI approximation than its QFH counterpart. Hence, we decided to use the semi-classical QFH approximation while using both the TI propagator as well as QFH* effective potential.

We know that all propagators yield results that converge to the correct value in the $P \to \infty$ limit, irrespective of the choice of the potential. So, as a first step, we verified that the $B_2^{\mathrm{sc,QFH^*}}$ did agree within statistical uncertainties with B_2^{cl}. In the next step, we break down our $B_2^{\mathrm{sc,QFH^*}}$ and $B_2^{\mathrm{sc,TI}}$ simulations into smaller, more precise ones using the decomposition algorithm. We observed a similar trend for $B_2^{\mathrm{sc,QFH^*}}$ and $B_2^{\mathrm{sc,TI}}$ decompositions as was observed [18] for B_2^{cl}, i.e. for $T > 63.15\,\mathrm{K}$ $[\Gamma^{\mathrm{SCL}}(u)]$ is always chosen as the preliminary approximation, for $4K \leq T \leq 63.15\,\mathrm{K}$ $[\Gamma^{\mathrm{SCL}}(u^{\mathrm{simple}})]$ is chosen as the preliminary approximation and for $T < 4\,\mathrm{K}$ $[\Gamma(P, u^{\mathrm{simple}})]$ is chosen as the preliminary approximation.

To assess the performance of SCB-QFH* and SCB-TI approaches against the CB approach in terms of achieving faster convergence as P increases, in Fig. 2, we plot the magnitude of $y = [\Gamma(P,u) - \Gamma(P/2,u)]$ as a function of P. For convergence to be achieved, as P increases $|y|$ decreases and as $P \to \infty$, $|y| \to 0$; the smaller the value of $|y|$, the faster the convergence. In Fig. 2 we see that the SCB-TI values are consistently lower than values of the other two approaches for all temperatures except T = 10.0 and 50.0 K for P = 4 beads, where the SCB-QFH* has lower $|y|$ values than SCB-TI. This condition is not particularly relevant, because at lower temperatures we almost always use a value of $P > 4$ and the convergence is more dependent on $|y|$ values for higher P (128 say), where SCB-TI has much lower $|y|$ values. From Fig. 2 we also notice that as temperature increases, $|y|$ decreases for each case. This is to be expected, because as we increase temperature the system approaches classical behavior, requiring fewer and fewer beads to converge.

To assess the performance of SCB-QFH* and SCB-TI approaches against the CB approach in terms of achieving better precision, we plot the ratios of uncertainty of the quantity $y = [\Gamma(P,u) - \Gamma(P/2,u)]$, i.e. we plot σ_y (SCB-QFH*)/σ_y(CB) and σ_y (SCB-TI)/σ_y(CB) in Fig. 3. In order to make a fair comparison, we use the uncertainties due to the same number of MC steps (1×10^6) for each case. In Fig. 3

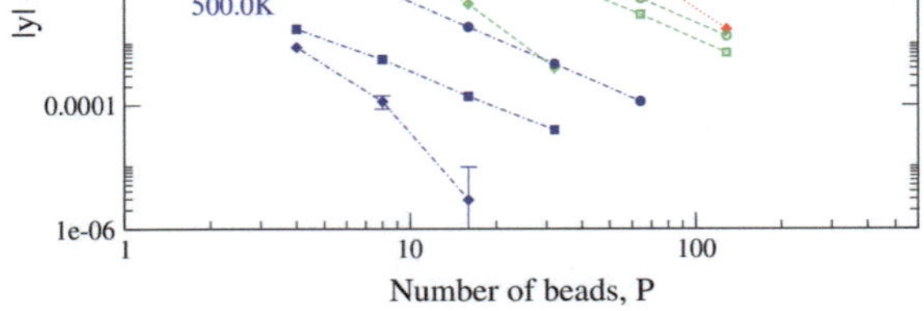

Fig. 2 Convergence factor, $y = [\Gamma(P,u) - \Gamma(P/2,u)]$ as a function of number of beads P. Symbols alternate filled or open with each temperature and indicate: classical-beads (CB) approach (*circles*); SCB-QFH* (*squares*); SCB-TI (*diamonds*). Temperatures are T = 2.5 K (*black open symbols* connected by *solid lines*); T = 10.0 K (*red filled symbols* connected by *dotted lines*); T = 50.0 K (*green open symbols* connected by *dashed lines*); T = 500.0 K (*blue filled symbols* connected by *dash-dot lines*). Confidence limits (68 %) are smaller than the symbol sizes except where shown

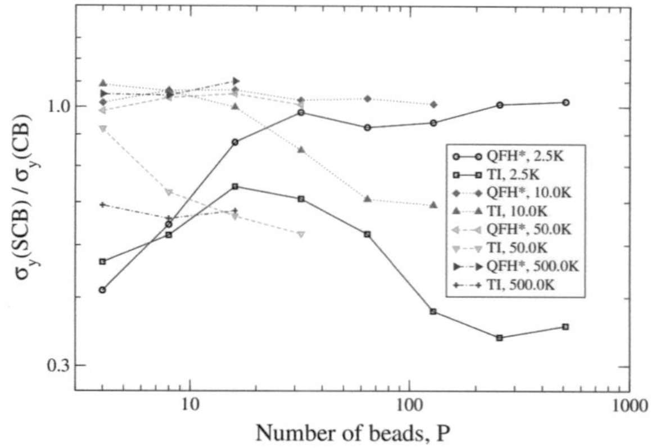

Fig. 3 Uncertainty ratio of the convergence factor $\sigma_y(\mathrm{SCB})/\sigma_y(\mathrm{CB})$, where $y = [\Gamma(P, u) - \Gamma(P/2, u)]$, as a function of number of beads P

we observe that SCB-TI has a consistently lower uncertainty ratio than SCB-QFH* for all P except for P = 4, where the T = 2.5 and 5.0 K results for SCB-QFH* have slightly lower values. At these low temperatures, since $P > 4$ almost always, we do not worry too much about SCB-QFH* having lower uncertainty ratios because it does not affect the uncertainty of the overall result that much, and also because the values are only slightly lower. The ratio for SCB-QFH* is almost always greater than 1, suggesting that it is not expected to give better precision when compared to CB for most cases. For the cases where the SCB-QFH* ratio is less than 1, i.e. T = 2.5 K and $P \leq 128$, we expect it to give better precision. The ratio for SCB-TI is almost always less than 1, suggesting that it is expected to give better precision when compared to CB for most cases. For the cases where the SCB-TI ratio is greater than 1, i.e. T = 10.0 K and $P \leq 8$, the magnitude is only marginally greater; as explained earlier, usually $P > 8$ is needed for accurate results when T = 10.0 K.

To assess the performance of SCB-QFH* and SCB-TI approaches against the CB approach in terms of the uncertainty achieved for a given period of time, in Fig. 4 we plot the ratios of the best-case uncertainties of the quantum second virial

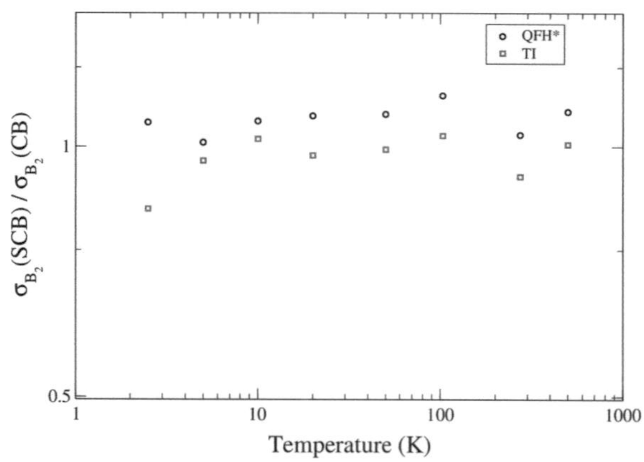

Fig. 4 Uncertainty ratio $\sigma_{B_2}(\mathrm{SCB})/\sigma_{B_2}(\mathrm{CB})$ as a function of temperature, for optimal decomposition with a fixed total computation time of 1 CPU-hour

coefficient values resulting from an overall simulation time of 1 h, i.e. we plot $\sigma_{B_2}(\text{SCB - QFH}^*)/\sigma_{B_2}(\text{CB})$ and $\sigma_{B_2}(\text{SCB} - \text{TI})/\sigma_{B_2}(\text{CB})$ calculated after optimally decomposing the simulation effort for a cumulative time of 1 h. We again observe that in Fig. 4, the SCB-TI approach has a lower uncertainty ratio than SCB-QFH* for all temperatures considered. Also, the ratio of SCB-TI is slightly less than 1 for most cases while that of SCB-QFH* is always greater than 1. This suggests that decomposition for the SCB-QFH* approach is expected to yield larger uncertainties for the quantum virial coefficient, compared to that of CB and SCB-TI approaches. The decomposition for the SCB-TI approach seems to be performing better than the SCB-QFH* approach, especially at lower temperatures, which is desirable because we normally tend to use large *P* at these temperatures. Even in the cases where the SCB-TI ratio is greater than 1, it is only marginally greater and therefore it may be considered acceptable.

6 Conclusion

We have implemented the PIMC method with two approaches based on semi-classical beads (SCB-QFH*, SCB-TI) and MSMC to compute more precise quantum virial coefficients for helium-4. The SCB results agree well with CB results as they are within statistical uncertainties of each other. The decomposition algorithm of Shaul et al. [18] was implemented to achieve better efficiency of quantum virial coefficient calculations. We observed similar trends in decompositions of simulations in our SCB based approaches as was the case for the CB approach. For lower temperatures, the approximation u^{simple} to u for finite P is chosen as the preliminary approximation. As the temperature increases, the preliminary approximation preferred is the semi-classical approximation to u^{simple}, and for high temperatures the semi-classical approximation to u is preferred. Having chosen the preliminary approximation, the decomposition algorithm spends the most time in the first step and the amount of time spent per step gradually decreases for subsequent steps. This is because the subsequent steps involve more computationally expensive calculations (either by shifting to u from its semi-classical approximation, or by doubling P from the previous step, or by shifting to the full potential u from u^{simple}) and by design, these steps also yield better and better precision. The decomposition algorithm was designed to allocate computational effort proportional to the difficulty of the computation, which is defined as ((cpu-time)$^{1/2}$ $\times$ uncertainty). The SCB-QFH* and SCB-TI approaches have comparable and better uncertainties respectively, for the steps that involve computing $[\Gamma(P, u) - \Gamma(P/2, u)]$ or $[\Gamma(P, u^{\text{simple}}) - \Gamma(P/2, u^{\text{simple}})]$. Since these steps involve significant computational costs and relatively low uncertainties, the amount of effort dedicated for them is lower, attenuating the effect of any efficiency brought to their calculation. As a result, the improvement of the precision of the resulting virial coefficient is only marginal. We note that if the decomposition algorithm is not being used, either because it is non-trivial to apply, or because virial coefficients are not being computed, the SCB-TI approach performs much better than both SCB-QFH* and CB approaches, which is what we would expect anyway from the use of a higher order propagator.

In summary, we found the following order for the rate of convergence with respect to number of beads *P*: SCB-TI > SCB-QFH* > CB. We expect a similar trend for the rate of convergence with respect to *P* for higher order coefficients as well, because of the use of the higher order TI propagator. The order for precision was found to be: SCB-TI > SCB-QFH*. Compared to CB, QFH* is always worse but only marginally so; TI is almost always better and only marginally worse for a few temperatures. We expected a trend similar to the rate of convergence with *P* for the precision as well, even for B_2 calculations. Since this was not what we observed, partially due to the decomposition algorithm, an understanding of the order of precision for higher order coefficients for the SCB based approaches compared to CB approach would require further investigation. However, we do expect the order between SCB based approaches to remain the same, i.e., SCB-TI > SCB-QFH*.

Directions for future work include investigating more temperatures, comparing the performance of different higher-order propagators of the thermal density matrix in terms of precision and rate of convergence, and using alternative ab initio potentials as they become available. Extension of PIMC with semi-classical beads to multi-atomic molecules is straightforward, and we expect such an approach to perform better than conventional PIMC with classical beads, in terms of convergence rate and precision.

Acknowledgments This work is supported by the U.S. National Science Foundation (CHE-1027963).

References

1. Feynman, R.P., Hibbs, A.R.: Quantum Mechanics and Path Integrals, 1st edn. McGraw-Hill Companies, Inc., New York (1965). Emended by Daniel F. Styer
2. Ceperley, D.: Path integrals in the theory of condensed helium. Rev. Mod. Phys. **67**, 279 (1995)
3. Cui, T., Cheng, E., Alder, B., Whaley, K.: Rotational ordering in solid deuterium and hydrogen: a path integral Monte Carlo study. Phys. Rev. B **55**, 12253 (1997)
4. Takahashi, M., Imada, M.: Monte Carlo calculation of quantum systems. II. higher order correction. J. Phys. Soc. Jpn. **53**, 3765–3769 (1984)
5. Schenter, G.K.: The development of effective classical potentials and the quantum statistical mechanical second virial coefficient of water. J. Chem. Phys. **117**, 6573 (2002)
6. Janke, W., Sauer, T.: Properties of higher-order Trotter formulas. Phys. Lett. A **165**, 199–205 (1992)
7. Suzuki, M.: Hybrid exponential product formulas for unbounded operators with possible applications to Monte Carlo simulations. Phys. Lett. A **201**, 425–428 (1995)
8. Yamamoto, T.M.: Path-integral virial estimator based on the scaling of fluctuation coordinates: application to quantum clusters with fourth-order propagators. J. Chem. Phys. **123**, 104101 (2005)

9. Garberoglio, G., Harvey, A.H.: First-principles calculation of the third virial coefficient of helium. J. Res. Natl. Inst. Stand. Technol. **114**, 249 (2009)
10. Fellmuth, B., Gaiser, C., Fischer, J.: Determination of the Boltzmann constant—status and prospects. Meas. Sci. Technol. **17**, R145–R159 (2006)
11. Schmidt, J.W., Gavioso, R.M., May, E.F., Moldover, M.R.: Polarizability of helium and gas metrology. Phys. Rev. Lett. **98**, 254504 (2007)
12. Pitre, L., Moldover, M.R., Tew, W.L.: Acoustic thermometry: new results from 273 K to 77 K and progress towards 4K. Metrologia **43**, 142–162 (2006)
13. Moldover, M.R., McLinden, M.O.: Using *ab initio* data to accurately determine the fourth density virial coefficient of helium. J. Chem. Thermodyn. **42**, 1193–1203 (2010)
14. Aziz, R.A., Janzen, A.R., Moldover, M.R.: *Ab initio* calculations for helium: a standard for transport property measurements. Phys. Rev. Lett. **74**, 1586–1589 (1995)
15. Shaul, K.R.S., Schultz, A.J., Kofke, D.A., Moldover, M.R.: Semiclassical fifth virial coefficients for improved *ab initio* helium-4 standards. Chem. Phys. Lett. **531**, 11–17 (2012)
16. Garberoglio, G., Harvey, A.H.: Path-integral calculation of the third virial coefficient of quantum gases at low temperatures. J. Chem. Phys. **134**, 134106 (2011)
17. Garberoglio, G., Moldover, M.R., Harvey, A.H.: Improved first-principles calculation of the third virial coefficient of helium. J. Res. Natl. Inst. Stand. Technol. **116**, 729–742 (2011)
18. Shaul, K.R., Schultz, A.J., Kofke, D.A.: Path-integral Mayer-sampling calculations of the quantum Boltzmann contribution to virial coefficients of helium-4. J. Chem. Phys. **137**, 184101 (2012)
19. Tester, J.W., Modell, M.: Thermodynamics and Its applications, 3rd edn. Prentice Hall Inc, New Jersey (1997)
20. Masters, A.J.: Virial expansions. J. Phys.: Condens. Matter **20**, 283102 (2008)
21. Hansen, J.P., McDonald, I.R.: Theory of Simple Liquids, 3rd edn. Academic Press (2006)
22. Przybytek, M., Cencek, W., Komasa, J., Łach, G., Jeziorski, B., Szalewicz, K.: Relativistic and quantum electrodynamics effects in the helium pair potential. Phys. Rev. Lett. **104**, 183003 (2010)
23. Singh, J.K., Kofke, D.A.: Mayer sampling: calculation of cluster integrals using free-energy perturbation methods. Phys. Rev. Lett. **92**, 220601 (2004)
24. Schultz, A.J., Kofke, D.A.: Sixth, seventh and eighth virial coefficients of the Lennard-Jones model. Mol. Phys. **107**, 2309 (2009)
25. Percus, J.K., Yevick, G.J.: Analysis of classical statistical mechanics by means of collective coordinates. Phys. Rev. **110**, 1–13 (1958)
26. Shaul, K.R.S., Schultz, A.J., Perera, A., Kofke, D.A.: Integral-equation theories and Mayer-sampling Monte Carlo: a tandem approach for computing virial coefficients of simple fluids. Mol. Phys. **109**, 2395–2406 (2011)

Homogeneous Nucleation of $[dmim^+][Cl^-]$ from its Supercooled Liquid Phase: A Molecular Simulation Study

Xiaoxia He, Yan Shen, Francisco R. Hung and Erik E. Santiso

Abstract We have used molecular simulations to study the homogeneous nucleation of the ionic liquid $[dmim^+][Cl^-]$ from its bulk supercooled liquid at 340 K. Our combination of methods include the string method in collective variables (Maragliano et al., J. Chem. Phys. 125:024106, 2006), Markovian milestoning with Voronoi tessellations (Maragliano et al J Chem Theory Comput 5:2589, 2009), and order parameters for molecular crystals (Santiso and Trout J Chem Phys 134:064109, 2011). The minimum free energy path, the approximate size of the critical nucleus, the free energy barrier and the rates involved in the homogeneous nucleation process were determined from our simulations. Our results suggest that the subcooled liquid (58 K of supercooling) has to overcome a free energy barrier of $\sim$85 kcal/mol, and has to form a critical nucleus of size $\sim$3.4 nm; this nucleus then grows to form the monoclinic crystal phase. A nucleation rate of 6.6×10^{10} cm^{-3} s^{-1} was determined from our calculations, which agrees with values observed in experiments and simulations of homogeneous nucleation of subcooled water.

Keywords Homogeneous nucleation · Ionic liquid · Molecular dynamics

---

X. He · Y. Shen · F.R. Hung (✉)
Cain Department of Chemical Engineering, Louisiana State University,
Baton Rouge, LA 70803, USA
e-mail: frhung@lsu.edu

F.R. Hung
Center for Computation & Technology, Louisiana State University, Baton Rouge,
LA 70803, USA

E.E. Santiso
Department of Chemical and Biomolecular Engineering, North Carolina State University,
Raleigh, NC 27695, USA

R.Q. Snurr et al. (eds.), *Foundations of Molecular Modeling and Simulation*,
Molecular Modeling and Simulation, DOI 10.1007/978-981-10-1128-3_7

1 Introduction

Room-temperature ionic liquids (ILs) have attracted significant attention as designer solvents, electrolytes, and other applications mostly involving liquid phases of the ILs. Very recently, Warner et al. [1–7] developed IL-based nanomaterials (dubbed GUMBOS, for Group of Uniform Materials Based on Organic Salts) where these compounds are in the solid state. These IL-based materials hold enormous promise, as they have the highly tunable properties of ILs [8, 9] and can be prepared via simple procedures [1–7], possibly impacting fields as diverse as optoelectronics, photovoltaics, separations, analytical chemistry and biomedicine. 1D-nanomaterials such as nanorods and nanowires were also synthesized [6] by introducing the ILs inside hard templates with cylindrical nanopores, e.g., multi-walled carbon nanotubes and anodic alumina membranes; shape anisotropy can lead to further variations in interesting properties of these nanomaterials (fluorescence, magnetic). On the other hand, ILs are also immobilized in nanoporous solids (carbon nanotubes, silica, cellulose, polymers, etc.) during the synthesis of ionogels [10]. These hybrid materials have potential applications in lithium batteries, fuel cells and solar cells, and in catalysis and biocatalysis, drug delivery and optical sensing devices [10]. A rational design of IL-based nanomaterials and ionogels require a fundamental understanding of the solidification, as well as the nucleation and growth of crystals of ILs in contact with surfaces and inside nanopores.

As a starting point for our studies in this area, here we focus on modeling the homogeneous nucleation of a simple IL, 1,3-dimethylimidazolium chloride, or [dmim$^+$][Cl$^-$], from its supercooled liquid phase in the bulk. This IL has been extensively studied in previous simulation reports [11–19]. However, nucleation is an extremely challenging problem [20–27], mainly because the initial stages of nucleation typically involve a few molecules or atoms, which makes it difficult to design experiments to study nucleation at the molecular level. Molecular dynamics (MD) simulations of nucleation are also very challenging, as nucleation is a rare event. Previous studies of homogeneous nucleation of systems of particles (hard-sphere, Lennard-Jones, etc.) [28–37] water [38–46] and other substances (e.g., NaCl, silicon, benzene, n-octane, urea, copper, aluminum, etc.) [47–58] using rare event methods have provided important insights on these phenomena in those systems. However, these methods might have limitations when studying the nucleation of ILs, which can have very slow dynamics and therefore the transition paths might take a prohibitively long time to commit to any of the stable states. Here we have combined the string method in collective variables (SMCV) [59] with Markovian Milestoning with Voronoi tessellations [60–62] to study the homogeneous nucleation of [dmim$^+$][Cl$^-$]. This combination of methods has been used before to study conformational changes in biomolecules [59, 61, 63, 64]. In the SMCV, a string of replicas connecting the liquid with the crystal phase evolves guided by the negative gradient of the free energy with respect to some collective variables (or order parameters, OPs, characterizing the transition), until they converge into a minimum free energy path (MFEP). This path represents the region where the transition has the maximum probability to take place. This converged string is then used as input to our second method, Markovian milestoning with Voronoi tessellations, which yields the free energy along the path, as well as the mean first passage times (MFPTs) in the transition studied. In association with these methods we have used the OPs for molecular crystals recently developed by Santiso and Trout [65]. These OPs can distinguish between different crystal polymorphs and liquid phases, and detect crystal ordering in nm-size regions. For example, if our system would crystallize into two solid polymorphs, one could use our methods to determine the minimum free energy paths connecting the supercooled liquid with the different polymorphs, and comparison of the free energies of the polymorphs would provide insights about their thermodynamic stability.

We note here that we have used the same methods and OPs in previous studies of the homogeneous nucleation of benzene [66] and the same IL studied here, [dmim$^+$][Cl$^-$] [67]. In the latter report we used a system consisting of 1372 ion pairs, and interpreted our results using calculations based on classical nucleation theory, which also helped us address possible finite-size effects in this system. Our interpretations were corroborated by SMCV results obtained using a larger system containing 2268 ion pairs [67]. Here in this paper we present a complete account and discussion of our results obtained for this larger system, using both the SMCV and Markovian milestoning with Voronoi tessellations, which completes the work we have presented before [67]. Nucleation could be studied using alternative approaches such as transition path sampling/aimless shooting, and metadynamics (see, e.g. [54, 55]). However, for very complex systems such as ILs, which have slow dynamics, the transition paths might take a long time to commit to any of the stable states, and trajectories could recross the top of the barrier multiple times [54], which could make sampling extremely challenging. Furthermore, and in order to avoid finite-size effects, here we had to consider a relatively large system size (2268 ion pairs), which required us to use a total of 180 OPs. Such a system would be extremely challenging to study using metadynamics (systems previously studied with this method were significantly smaller and had a considerably lower number of collective variables, see, e.g. [55, 68]). The rest of the paper is structured as follows. In the next section, we provide details of our systems and methods (OPs used, SMCV and Markovian milestoning with Voronoi tessellations). Section 3 contains our main results and discussion, and our main findings are summarized in Sect. 4.

2 Simulation Details

2.1 Models

The monoclinic crystal structure of [dmim$^+$][Cl$^-$] was obtained from the Cambridge Crystallographic Data Centre [69, 70]. The classical, non-polarizable, all-atom force field developed by Lopes et al. [71–75] was used to model the IL, mainly because it can reproduce the experimental crystal structure of [dmim$^+$][Cl$^-$] [71] (here the cations and anions have integer charges). Furthermore, the density of the liquid phase and the melting point as determined from simulations using this model were found to be in good agreement with experimental values [76]. Our system contained 2268 ion pairs, which can form a monoclinic crystal of characteristic dimensions 8.3 nm × 5.3 nm × 9.7 nm. All MD simulations were performed using a modified version of the NAMD software [77] that included implementations of the OPs, the SMCV and Markovian milestoning with Voronoi tessellations (see below) in C++ libraries. All the simulations were performed in the NPT ensemble with $P = 1$ bar and $T = 340$ K (a supercooling of 58 K). When selecting the temperature (and thus the degree of supercooling) in our systems, we considered the following aspects that would impact the computational costs of our simulations. Higher temperatures (i.e., smaller supercoolings) would result in an increase in the size of the critical nucleus. In this situation one could obtain a critical nucleus that might be larger than the dimensions of the simulation box, and run into finite-size issues; therefore a larger simulation box would be needed, which would increase computational costs. Reducing the temperature (i.e., having a larger supercooling) would reduce the size of the critical nucleus; however, if the temperature is too low, the dynamics of the ions would slow down and thus very long simulations would be needed. We have found that, for our system, 58 K of supercooling provides a system with reasonable dynamics, and allowed us to use a simulation box of reasonable size (so that the critical nucleus formed does not percolate through any of the dimensions of the box). A Langevin thermostat with a damping coefficient of 25 ps^{-1} was used to control the temperature, and the Nosé-Hoover Langevin piston with a damping time of 50 fs was used as the barostat. Periodic boundary conditions were applied in all directions. Lennard-Jones and electrostatic interactions were cutoff at 10 and 12 Å; particle mesh Ewald (PME) [78] was used to handle the latter type of interactions. Hydrogen bond lengths were constrained with the LINCS algorithm, and a time step of 0.5 fs was used in our simulation runs [79]. Additional details of our model systems can be found elsewhere [67].

2.2 *Order Parameters (OPs)*

The OPs developed by Santiso and Trout [65] are extracted from a generalized pair distribution function. All OPs used in this study were based on [dmim$^+$]; no particular OPs were defined for [Cl$^-$]. In Fig. 1 two ion pairs are shown, where the absolute orientation of each cation is given by the vectors q_1 and q_2, which are normal to the imidazolium ring of each cation. The distance OP provides quantification for the various center of mass (COM) distances between the cations. The bond orientation OP measures the orientation of bonds joining the center-of-mass of the cations, while the relative orientation OP measures the orientation of one cation with respect to another one [65, 67]. All the OPs are defined per cation and per peak

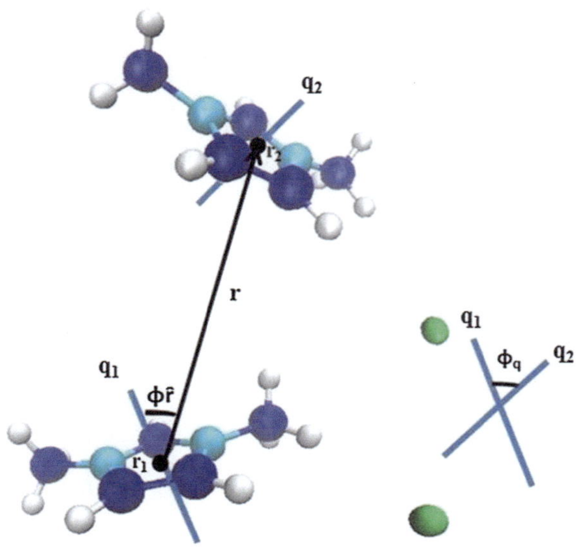

Fig. 1 Variables used in the construction of OPs. The vector normal to the plane of the imidazolium ring of [dmim$^+$] gives the absolute orientation of each of the two cations. The distance OP is based on r, which joins the center of mass (COM) of the two cations. The angle formed by the vectors r and q_1 is used for the bond orientation $\phi_{\hat{r}}$, whereas the angle formed by the vectors q_1 and q_2 is used for the relative orientation ϕ_q

in the pair distribution function, which leads to an extremely large number of OPs. To reduce the total number of OPs, we first sum over peaks in the pair distribution function, and then calculate local averages over the cations present in a given subcell of the simulation box, according to the following equation [65]:

$$\theta_C = \frac{1}{N_C}\sum_{i\in C}\sum_{\alpha}\varphi_{i,\alpha} \tag{1}$$

where N_C is the number of cations in subcell C, the index i denotes the ith cation, the index α runs over peaks of the pair distribution function, and $\varphi_{i,\alpha}$ represents any of the per-molecule and per-peak OP (i.e., distance, bond orientation or relative orientation). Therefore, each subcell has a corresponding OP. Additional details are presented elsewhere [65].

In our system, the simulation box was divided into $6 \times 5 \times 6$ subcells, giving a total of 180 OPs of each type (distance, bond orientation and relative orientation). In Fig. 2 we present the number frequency of distance, bond orientation and relative orientation OPs for the liquid and crystal phases of [dmim$^+$][Cl$^-$] at 340 K. The frequency is averaged over 400 configurations (as obtained from short, 0.4 ns MD simulation of the liquid and solid phases in the NPT ensemble). These results indicate that any of the OPs can serve as a good metric to distinguish between crystal and liquid phases, even in nm-sized regions. Here we chose to work with the bond orientation OPs, although we also monitored the distance and the relative orientation OPs.

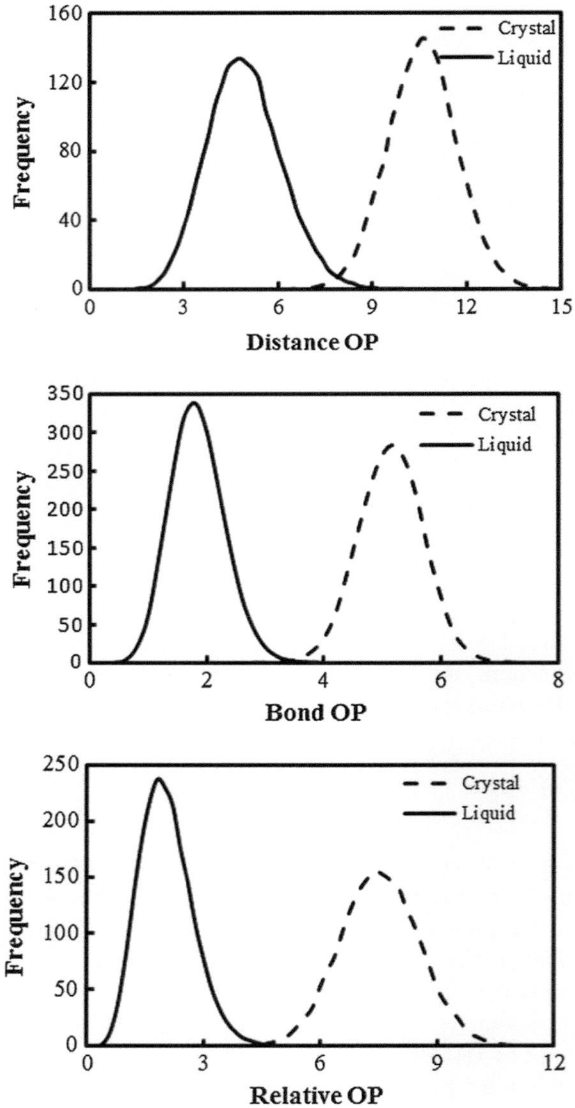

Fig. 2 The number frequency of distance, bond orientation and relative orientation OPs for liquid and solid IL, as obtained from 400 liquid-like and 400 crystal-like configurations of the IL. The distance, bond orientation and relative orientation OPs have units of $Å^{-1}$, which corresponds to the units of σ_α (see, e.g., Eqs. 11, 15, 21–22 in Ref. [65]; we used similar OPs here)

2.3 String Method in Collective Variables (SMCV)

The SMCV [59, 63] was used to sketch a MFEP for the homogeneous nucleation of $[dmim^+][Cl^-]$ from its subcooled liquid. Here we used the following procedure:

1. An initial string consisting of 32 replicas was prepared by collecting a number of intermediate states from the simulated melting of a crystal of the IL at 800 K.

Each replica i in this string contains 180 OPs (denoted by q_i) that characterize its local structure.

2. At every step of the SMCV method:

   2.1. An extended Hamiltonian is established for each replica, by including a set of harmonic springs that keep each replica's OPs close to their 'target' values. Using this extended Hamiltonian, we run short (0.2 ns) MD simulations in the NPT ensemble for each image, in order to determine the mean force $\nabla_q F(q_i)$ and metric tensor $M(q_i)$ required to maintain each image close to the target OP values (here the mean force is the negative gradient of the free energy with respect to the OPs)

   2.2. The new target OPs are estimated by taking a forward Euler step on the string evolution equation:

$$q_i^* = q_i - \Delta\tau M(q_i)\nabla_q F(q_i) \tag{2}$$

   where $\Delta\tau$ is the time step in the SMCV, and q_i^* denotes the target OPs for the ith image for the next SMCV step.

   2.3. Reparameterize by interpolating a curve through the new target OP values q_i^*, and recompute new target OP values q_i^* so that consecutive replicas are at constant arclength from each other. This step prevents the replicas from clustering near the stable states.

3. Step 2.1–2.3 above are repeated until the potential of mean force (PMF) converges; here the PMF is the line integral of the restraint force along the path in the multi-dimensional OP space

Additional details about the SMCV and its implementation are presented elsewhere [59, 63, 66, 67].

2.4 Markovian Milestoning with Voronoi Tessellations

The MFEP computed from the SMCV is an n-dimensional function (where n is equal to the 180 OPs), determined by effectively restraining 180 degrees of freedom as the replicas in the SMCV evolve following the negative gradient of the free energy with respect to the OPs. The MFEP thus only represents a single pathway in the transition tube. However, this tube can be mapped into a single free energy curve as a function of a reaction coordinate, by using the converged string from the SMCV as input to simulations using Markovian milestoning with Voronoi tessellations [60–63]. This procedure can also yield information about the rate of nucleation. In this method, we associate a cell in OP space to each replica from the converged SMCV string, forming a tessellation in the OP space. By construction [61, 63], the edges of the tessellations are approximate isocommittor surfaces, and thus are used as milestones in our procedure. We then perform MD simulations where each replica is forced to remain within its own cell by using soft walls (planar half-pseudoharmonic restraints) [61]. In these simulations we monitor the number of collisions each trajectory does with the milestones (edges of the Voronoi tessellation). At steady state, if $N_{n,m}$ represents the number of collisions that the MD trajectory in cell B_n experiences with the boundary of cell B_m during the simulation time t_n, the rate of escape from tessellation cell B_n to cell B_m can be estimated as:

$$v_{n,m} = \frac{N_{n,m}}{t_n} \tag{3}$$

The probability π_n of finding the system in cell B_n can then be calculated from the following equations:

$$\sum_{\substack{n=1 \\ n \neq m}}^{N} \pi_n v_{n,m} = \sum_{\substack{n=1 \\ n \neq m}}^{N} \pi_m v_{m,n} \tag{4}$$

$$\sum_{n=1}^{N} \pi_n = 1 \tag{5}$$

The free energy curve F_n can be calculated as a function of cell n as follows:

$$F_n = -k_{\mathrm{B}} T \ln \pi_n \tag{6}$$

In turn, the MFPTs can be also computed from the Voronoi milestoning procedure using the following equations:

$$\sum_{b \neq b^*} k_{a,b}\, t_{b,b^*} = -1, \quad a \neq b^* \tag{7}$$

$$k_{a,b} = \frac{\sum_{n=0}^{N} \pi_n N_{a,b}^{n} \big/ t_n}{\sum_{n=0}^{N} \pi_n t_a^n \big/ t_n} \tag{8}$$

where $t_{b,b*}$ is the MFPT to milestone b^* from other milestones in the system (where $b \neq b^*$). $k_{a,b}$ is the rate of instantaneous transition from milestone a to milestone b; $N_{a,b}^n$ is the total number of transitions from milestone a to milestone b during the simulation confined to cell B_n, and t_a^n is total time during which a was the most recent milestone visited by the system. We used the arbitrary precision floating-point library implemented in the Sage software [80] to solve the equations above, as the values of $k_{a,b}$ could have variations of up to 6 orders of magnitude. Additional details of these simulations are provided elsewhere [61, 63, 66, 67].

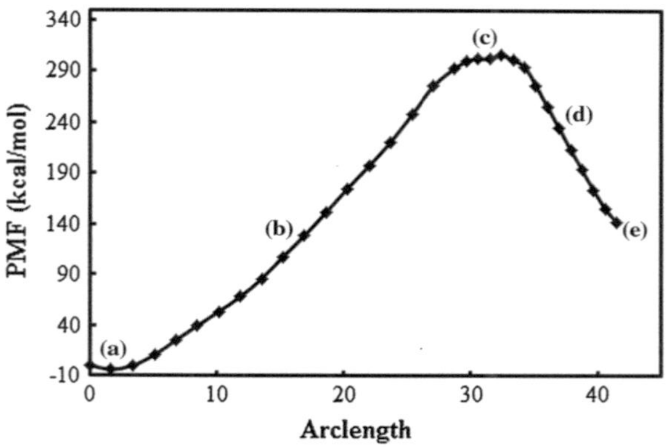

Fig. 3 The PMF associated with the MFEP for homogeneous nucleation of a crystal phase of [dmim$^+$][Cl$^-$] from its supercooled liquid phase at 340 K and 1 bar. The *left* and *right sides of the curve* correspond to crystal and liquid states. Simulation snapshots of states labeled here are shown in Fig. 4

3 Results and Discussion

3.1 Determination of the MFEP from the SMCV

The converged PMF (see Sect. 2.3) as determined from our simulations with the SMCV is presented in Fig. 3; here the arclength (of the bond orientation OPs, in this case) represents the distance along the multidimensional nucleation path. A difference of about 141 kcal/mol is observed between the crystal phase and the supercooled liquid phase at about 58 K of supercooling, with a PMF barrier of about 163 kcal/mol between the supercooled liquid and the state at the top of the PMF profile.

In Fig. 4 we show *x–y* side views of representative simulation snapshots of several relevant states along the MFEP mapped in Fig. 3. As we move left from the supercooled liquid (state e) at the right of the PMF curve shown in Fig. 3, the cations and anions start to rearrange (state d) until we reach the top of the barrier (state c), where the ions form a critical nucleus exhibiting crystal-like order. If we arbitrarily define the cations with OPs > 3.5 (Fig. 2) as crystalline, those cations form a cluster in the replica at the top of the PMF curve (the critical nucleus) which has an average size of ~3.4 nm at this degree of supercooling. A similar procedure was used for the configuration at the top of the free energy curve (Fig. 6) determined from our Voronoi milestoning simulations (see below), giving a similar average size for the critical nucleus. Moving further left (and now downhill) along the MFEP mapped in Fig. 3, the crystal-like region grows (state b) until the system crystallizes completely (state a).

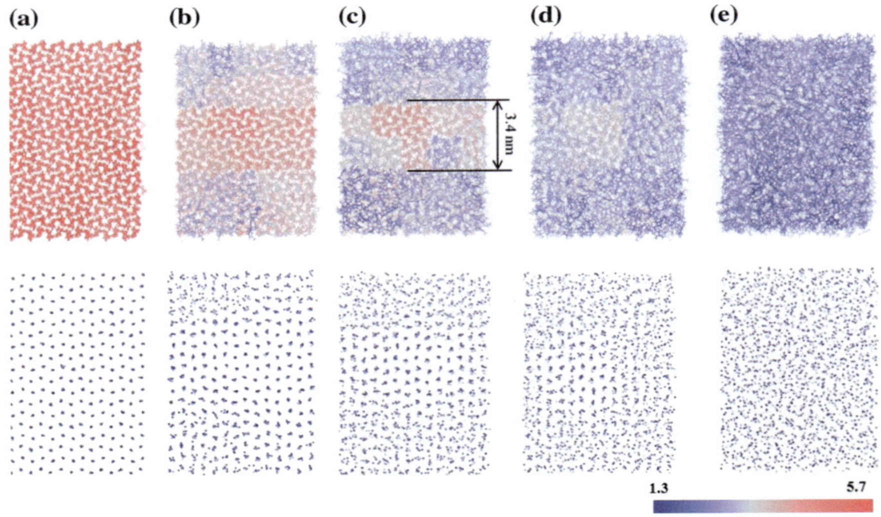

Fig. 4 x–y side views of simulation snapshots of states labeled along the MFEP shown in Fig. 3. The cations (*top row*) are *color-coded* according to the value of their bond orientation OPs (*red* crystal-like, *blue* liquid-like). The anions are shown on *bottom row* and are *not color-coded*

3.2 *Free Energy and MFPTs from Markovian Milestoning with Voronoi Tessellations*

The 32 replicas from the converged SMCV string were used as the starting point for our series of simulations with Markovian milestoning. Here we had to interpolate additional replicas (to reach a total of 74) as to ensure collection of enough statistics for the numerical solution of Eqs. (3–8). As direct analysis of the data is difficult due to the 180-dimensionality of the OP space, we used principal component analysis to determine the subspace of the OP space that contains most of the variance along the solidification path. In Fig. 5 we show the projections onto the first two principal components of the initial configurations (converged SMCV string, black dots), as well as some representative configurations (different colors) observed throughout our Markovian milestoning simulations. The results shown in Fig. 5 suggest that the images mainly remain within their own cells, although they occasionally wander into neighboring cells for a brief period; this is a consequence of using soft walls [61] to maintain each MD trajectory within its own cell. One important challenge is to properly distribute the images along the MFEP, in order to accumulate good statistics for the numerical solution of Eqs. (3–8). At the same time, we strived to not place too many images in MFEP regions with high curvature; this way, replicas mostly visit adjacent cells when they leave their own cells, and we avoid transitions between non-adjacent cells as much as possible (as discussed before [63], replicas visiting cells of non-adjacent neighbors can affect the accuracy in the calculations of the free energy and the MFPTs).

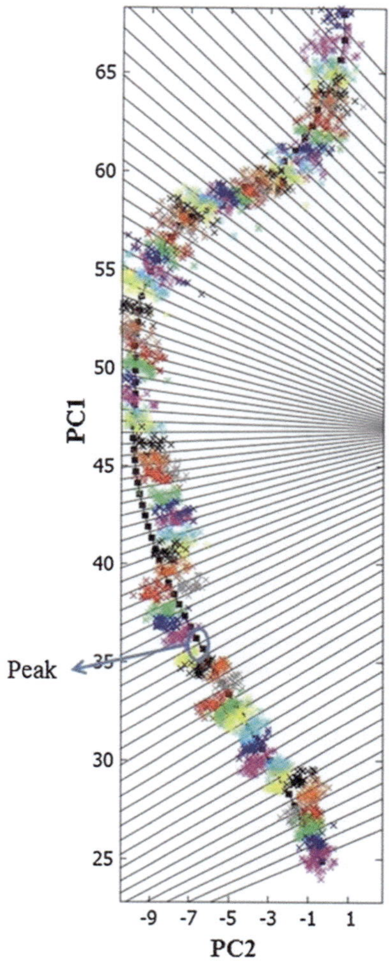

Fig. 5 Voronoi tessellation of the MFEP as projected onto the first two principal components PC1 and PC2 in OP space. The *black dots* represent the initial configurations (converged SMCV string plus interpolation of additional replicas) used in the Markovian milestoning simulations. Projections of representative configurations obtained from the milestoning procedure are also shown using *different colors*. The region corresponding to the peak of the PMF and free energy curves (Figs. 4 and 6) is labeled

In Fig. 6 we show the free energy along the MFEP as a function of the Voronoi cell number, as determined from the Voronoi milestoning simulations. These results indicate that the difference in free energy between the crystal and liquid phases of [dmim$^+$][Cl$^-$] at 58 K of supercooling is about 50 kcal/mol, and the free energy barrier between the liquid and the configuration at the top of the curve is about 85 kcal/mol. These free energy differences are comparable to those determined for water [41] and urea [55] in recent simulation studies of homogeneous nucleation. The free energy curve (Fig. 6) and the PMF curve (Fig. 3) are qualitatively similar (the snapshots of configurations obtained from the Markovian milestoning simulation procedure look very similar to those obtained from the SMCV and shown in Fig. 4); however the differences in free energies between relevant states are smaller than the corresponding differences in PMF. This observation is expected, as in the

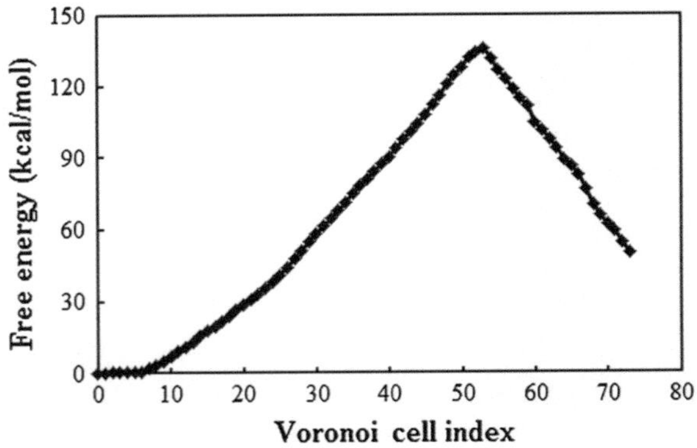

Fig. 6 Free energy involved in the homogeneous nucleation of $[dmim^+][Cl^-]$ from its supercooled liquid phase at 340 K and 1 bar, as obtained from the milestoning procedure. The *left* and *right sides of the curve* correspond to crystal and liquid states

SMCV more entropy is removed as we are effectively restraining 180 degrees of freedom (replicas evolve guided by the negative gradient of free energy in order parameter space). In contrast, in the Markovian milestoning simulations less entropy is removed from the system, as by construction a trajectory restrained to remain in its Voronoi cell is equal to the equivalent sections of a conventional (unbiased) MD trajectory that is passing through the same cell [61].

In Fig. 7 we show the MFPTs to reach the configurations at the peak of the free energy curve (Fig. 6). Here we only considered fluxes between adjacent cells (fluxes between non-adjacent cells represent about 13 % of the total number of fluxes in our milestoning simulations). If we consider all fluxes (i.e. between nearest and non-nearest neighboring images), we determined that the difference in MFPTs with respect to those shown in Fig. 7 is only of about 8 %. These results suggest that the contribution of non-adjacent isocommittor surfaces to the kinetics of the nucleation process can be ignored (for a detailed discussion, see Appendix B in the study of Ovchinnikov et al. [63]). The results shown in Fig. 7 suggest that the MFPTs to the configurations at the peak of the free energy curve of Fig. 6 are approximately constant for milestones $B_{i+1} \cap B_i$ with $i = 72, 71, \ldots, 53$. From the data shown in Fig. 7, we calculated a simulated nucleation rate of 6.6×10^{10} cm^{-3} s^{-1} for our system, which is at a supercooling of 58 K. Unfortunately no experimental data is available for our system; however, our computed rate is in

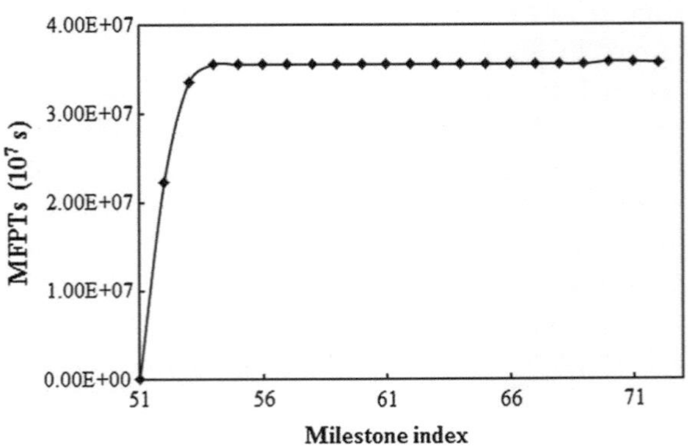

Fig. 7 Mean first passage times (MFPTs) from the milestone $B_{i+1} \cap B_i$ for $i = 72, 71, \ldots 52$ to the milestones at the shoulder of the free energy curve, $B_{52} \cap B_{51}$

reasonable agreement with experimental and simulation values for the homogeneous nucleation of ice, which are on the order of 10^{10} cm^{-3} s^{-1} [41, 81, 82] at a supercooling of about 40 K.

4 Conclusions

The homogeneous nucleation of the IL [dmim$^+$][Cl$^-$] from its bulk subcooled liquid phase (58 K of supercooling) was studied using molecular simulation. The SMCV [59, 63] combined with OPs [65] for molecular crystals was used to have a string of replicas map a MFEP connecting the supercooled liquid with the monoclinic crystal phase. The converged SMCV string was then used to initiate simulations using Markovian milestoning with Voronoi tessellations [61, 63]. These methods yield information about the free energy barrier, the size of the critical nucleus and the rate of nucleation. Our results indicate that the supercooled liquid has to overcome a free energy barrier of ~85 kcal/mol to form a critical nucleus of size ~3.4 nm. A simulated homogeneous nucleation rate of 6.6×10^{10} cm^{-3} s^{-1} was determined from our calculations. The values of the free energy barrier and the rate of nucleation are in reasonable agreement with experimental and simulation values obtained for the homogeneous nucleation of water and urea. Current work in our group is focused on the study of nucleation of ILs near surfaces and inside pores.

Acknowledgments We are grateful to Isiah Warner and his group (Chemistry, LSU) for helpful discussions. This work was partially supported by the National Science Foundation (CAREER Award CBET-1253075, and EPSCoR Cooperative Agreement EPS-1003897), and by the Louisiana Board of Regents. High-performance computational resources for this research were provided by High Performance Computing at Louisiana State University (http://www.hpc.lsu.edu) and by the Louisiana Optical Network Initiative (http://www.loni.org).

References

1. Tesfai, A., El-Zahab, B., Bwambok, D.K., Baker, G.A., Fakayode, S.O., Lowry, M., Warner, I.M.: Controllable formation of ionic liquid micro- and nanoparticles via a melt-emulsion-quench approach. Nano Lett. **8**, 897–901 (2008)
2. Tesfai, A., El-Zahab, B., Kelley, A.T., Li, M., Garno, J.C., Baker, G.A., Warner, I.M.: Magnetic and nonmagnetic nanoparticles from a group of uniform materials based on organic salts. ACS Nano **3**, 3244–3250 (2009)
3. Bwambok, D.K., El-Zahab, B., Challa, S.K., Li, M., Chandler, L., Baker, G.A., Warner, I.M.: Near-Infrared fluorescent NanoGUMBOS for biomedical imaging. ACS Nano **3**, 3854–3860 (2009)
4. Das, S., Bwambok, D., El-Zahab, B., Monk, J., de Rooy, S.L., Challa, S., Li, M., Hung, F.R., Baker, G.A., Warner, I.M.: Nontemplated approach to tuning the spectral properties of cyanine-based fluorescent nanogumbos. Langmuir **26**, 12867–12876 (2010)

5. Dumke, J.C., El-Zahab, B., Challa, S., Das, S., Chandler, L., Tolocka, M., Hayes, D.J., Warner, I.M.: Lanthanide-based luminescent NanoGUMBOS. Langmuir **26**, 15599–15603 (2010)
6. de Rooy, S.L., El-Zahab, B., Li, M., Das, S., Broering, E., Chandler, L., Warner, I.M.: Fluorescent one-dimensional nanostructures from a group of uniform materials based on organic salts. Chem. Commun. **47**, 8916–8918 (2011)
7. Warner, I.M., El-Zahab, B., Siraj, N.: Perspectives on moving ionic liquid chemistry into the solid phase. Anal. Chem. **86**, 7184–7191 (2014)
8. Thomas, W.P.W.: Ionic Liquids in Synthesis. Wilet-VCH, Weinheim (2008)
9. Plechkova, N.V., Seddon, K.R.: Applications of ionic liquids in the chemical industry. Chem. Soc. Rev. **37**, 123–150 (2008)
10. Le Bideau, J., Viau, L., Vioux, A.: Ionogels, ionic liquid based hybrid materials. Chem. Soc. Rev. **40**, 907–925 (2011)
11. Sha, M., Wu, G., Fang, H., Zhu, G., Liu, Y.: Liquid-to-solid phase transition of a 1,3-dimethylimidazolium chloride ionic liquid monolayer confined between graphite walls. J. Phys. Chem. C **112**, 18584–18587 (2008)
12. Sha, M., Wu, G., Liu, Y., Tang, Z., Fang, H.: Drastic phase transition in ionic liquid Dmim CI confined between graphite walls: new phase formation. J. Phys. Chem. C **113**, 4618–4622 (2009)
13. Pinilla, C., Del Popolo, M.G., Kohanoff, J., Lynden-Bell, R.M.: Polarization relaxation in an ionic liquid confined between electrified walls. J. Phys. Chem. B **111**, 4877–4884 (2007)
14. Pinilla, C., Del Popolo, M.G., Lynden-Bell, R.M., Kohanoff, J.: Structure and dynamics of a confined ionic liquid. topics of relevance to dye-sensitized solar cells. J. Phys. Chem. B **109**, 17922–17927 (2005)
15. Youngs, T.G.A., Hardacre, C.: Application of static charge transfer within an ionic-liquid force field and its effect on structure and dynamics. ChemPhysChem **9**, 1548–1558 (2008)
16. Hanke, C.G., Atamas, N.A., Lynden-Bell, R.M.: Solvation of small molecules in imidazolium ionic liquids: a simulation study. Green Chem. **4**, 107–111 (2002)
17. Del Popolo, M.G., Lynden-Bell, R.M., Kohanoff, J.: Ab initio molecular dynamics simulation of a room temperature ionic liquid. J. Phys. Chem. B **109**, 5895–5902 (2005)
18. Buhl, M., Chaumont, A., Schurhammer, R., Wipff, G.: Ab initio molecular dynamics of liquid 1,3-dimethylimidazolium chloride. J. Phys. Chem. B **109**, 18591–18599 (2005)
19. Monk, J., Singh, R., Hung, F.R.: Effects of Pore size and pore loading on the properties of ionic liquids confined inside nanoporous CMK-3 carbon materials. J. Phys. Chem. C **115**, 3034–3042 (2011)
20. Debenedetti, P.G.: Metastable Liquids: Concepts and Principles. Princeton University Press, Princeton, NJ (1996)
21. Kaschiev, D.: Nucleation: Basic Theory with Applications. Butterworth-Heinemann, Oxford (2000)
22. Price, S.L.: Computed crystal energy landscapes for understanding and predicting organic crystal structures and polymorphism. Acc. Chem. Res. **42**, 117–126 (2008)
23. Erdemir, D., Lee, A.Y., Myerson, A.S.: Nucleation of crystals from solution: classical and two-step models. Acc. Chem. Res. **42**, 621–629 (2009)
24. Vekilov, P.G.: Nucleation. Cryst. Growth Des. **10**, 5007–5019 (2010)
25. Auer, S., Frenkel, D.: Quantitative prediction of crystal-nucleation rates for spherical colloids: a computational approach. Annu. Rev. Phys. Chem. **55**, 333–361 (2004)
26. Anwar, J., Zahn, D.: Uncovering molecular processes in crystal nucleation and growth by using molecular simulation. Angewandte Chemie-International Edition **50**, 1996–2013 (2011)
27. Palmer, J.C., Debenedetti, P.G.: Recent advances in molecular simulation: a chemical engineering perspective. AIChE J. **61**, 370–383 (2015)
28. TenWolde, P.R., RuizMontero, M.J., Frenkel, D.: Numerical calculation of the rate of crystal nucleation in a Lennard-Jones system at moderate undercooling. J. Chem. Phys. **104**, 9932–9947 (1996)

29. Vehkamäki, H., Ford, I.J.: Critical cluster size and droplet nucleation rate from growth and decay simulations of Lennard-Jones clusters. J. Chem. Phys. **112**, 4193–4202 (2000)
30. Auer, S., Frenkel, D.: Prediction of absolute crystal-nucleation rate in hard-sphere colloids. Nature **409**, 1020–1023 (2001)
31. Moroni, D., ten Wolde, P.R., Bolhuis, P.G.: Interplay between structure and size in a critical crystal nucleus. Phys. Rev. Lett. **94**, 235703 (2005)
32. Trudu, F., Donadio, D., Parrinello, M.: Freezing of a Lennard-Jones fluid: from nucleation to spinodal regime. Phys. Rev. Lett. **97**, 105701 (2006)
33. Desgranges, C., Delhommelle, J.: Insights into the molecular mechanism underlying polymorph selection. J. Am. Chem. Soc. **128**, 15104–15105 (2006)
34. Desgranges, C., Delhommelle, J.: Polymorph selection during the crystallization of Yukawa systems. J. Chem. Phys. **126**, 054501 (2007)
35. Jungblut, S., Dellago, C.: Heterogeneous crystallization on tiny clusters. EPL (Europhysics Letters) **96**, 56006 (2011)
36. Beckham, G.T., Peters, B.: Optimizing nucleus size metrics for liquid-solid nucleation from transition paths of near-nanosecond duration. J. Phys. Chem. Lett. **2**, 1133–1138 (2011)
37. Chkonia, G., Wölk, J., Strey, R., Wedekind, J., Reguera, D.: Evaluating nucleation rates in direct simulations. J. Chem. Phys. **130**, 064505 (2009)
38. Radhakrishnan, R., Trout, B.L.: Nucleation of hexagonal ice (Ih) in liquid water. J. Am. Chem. Soc. **125**, 7743–7747 (2003)
39. Li, T., Donadio, D., Russo, G., Galli, G.: Homogeneous ice nucleation from supercooled water. Phys. Chem. Chem. Phys. **13**, 19807–19813 (2011)
40. Reinhardt, A., Doye, J.P.K.: Free energy landscapes for homogeneous nucleation of ice for a monatomic water model. J. Chem. Phys. **136**, 054501 (2012)
41. Sanz, E., Vega, C., Espinosa, J.R., Caballero-Bernal, R., Abascal, J.L.F., Valeriani, C.: Homogeneous ice nucleation at moderate supercooling from molecular simulation. J. Am. Chem. Soc. **135**, 15008–15017 (2013)
42. Sear, R.P.: The non-classical nucleation of crystals: microscopic mechanisms and applications to molecular crystals, ice and calcium carbonate. Int. Mater. Rev. **57**, 328–356 (2012)
43. Andrey, V.B., Jamshed, A., Ruslan, D., Richard, H.: Challenges in molecular simulation of homogeneous ice nucleation. J. Phys.: Condens. Matter **20**, 494243 (2008)
44. Reinhardt, A., Doye, J.P.K.: Note: homogeneous TIP4P/2005 ice nucleation at low supercooling. J. Chem. Phys. **139**, 096102 (2013)
45. Joswiak, M.N., Duff, N., Doherty, M.F., Peters, B.: Size-dependent surface free energy and tolman-corrected droplet nucleation of TIP4P/2005 water. J. Phys. Chem. Lett. **4**, 4267–4272 (2013)
46. Holten, V., Limmer, D.T., Molinero, V., Anisimov, M.A.: Nature of the anomalies in the supercooled liquid state of the mW model of water. J. Chem. Phys. **138**, 174501 (2013)
47. Valeriani, C., Sanz, E., Frenkel, D.: Rate of homogeneous crystal nucleation in molten NaCl. J. Chem. Phys. **122** (2005)
48. Quigley, D., Rodger, P.M.: Free energy and structure of calcium carbonate nanoparticles during early stages of crystallization. J. Chem. Phys. **128**, 221101 (2008)
49. Li, T., Donadio, D., Galli, G.: Nucleation of tetrahedral solids: a molecular dynamics study of supercooled liquid silicon. J. Chem. Phys. **131**, 224519 (2009)
50. Yi, P., Rutledge, G.C.: Molecular simulation of crystal nucleation in n-octane melts. J. Chem. Phys. **131**, 134902 (2009)
51. Saika-Voivod, I., Poole, P.H., Bowles, R.K.: Test of classical nucleation theory on deeply supercooled high-pressure simulated silica. J. Chem. Phys. **124**, 224709 (2006)
52. Agarwal, V., Peters, B.: Nucleation near the eutectic point in a Potts-lattice gas model. J. Chem. Phys. **140**, 084111 (2014)
53. Singh, M., Dhabal, D., Nguyen, A.H., Molinero, V., Chakravarty, C.: Triplet correlations dominate the transition from simple to tetrahedral liquids. Phys. Rev. Lett. **112**, 147801 (2014)
54. Shah, M., Santiso, E.E., Trout, B.L.: Computer simulations of homogeneous nucleation of benzene from the melt. J. Phys. Chem. B **115**, 10400–10412 (2011)

55. Giberti, F., Salvalaglio, M., Mazzotti, M., Parrinello, M.: Insight into the nucleation of urea crystals from the melt. Chem. Eng. Sci. **121**, 51–59 (2015)
56. Yu, T.-Q., Chen, P.-Y., Chen, M., Samanta, A., Vanden-Eijnden, E., Tuckerman, M.: Order-parameter-aided temperature-accelerated sampling for the exploration of crystal polymorphism and solid-liquid phase transitions. J. Chem. Phys. **140**, 214109 (2014)
57. Samanta, A., Tuckerman, M.E., Yu, T.-Q.: E, W. Microscopic mechanisms of equilibrium melting of a solid. Science **346**, 729–732 (2014)
58. Pedersen, U.R., Hummel, F., Dellago, C.: Computing the crystal growth rate by the interface pinning method. J. Chem. Phys. **142**, 044104 (2015)
59. Maragliano, L., Fischer, A., Vanden-Eijnden, E., Ciccotti, G.: String method in collective variables: minimum free energy paths and isocommittor surfaces. J. Chem. Phys. **125**, 024106 (2006)
60. Vanden-Eijnden, E., Venturoli, M.: Revisiting the finite temperature string method for the calculation of reaction tubes and free energies. J. Chem. Phys. **130**, 194103 (2009)
61. Maragliano, L., Vanden-Eijnden, E., Roux, B.: Free energy and kinetics of conformational transitions from voronoi tessellated milestoning with restraining potentials. J. Chem. Theory Comput. **5**, 2589–2594 (2009)
62. Vanden-Eijnden, E., Venturoli, M.: Markovian milestoning with Voronoi tessellations. J. Chem. Phys. **130**, 194101 (2009)
63. Ovchinnikov, V., Karplus, M., Vanden-Eijnden, E.: Free energy of conformational transition paths in biomolecules: the string method and its application to myosin VI. J. Chem. Phys. **134**, 085103 (2011)
64. Miller, T.F., III; Vanden-Eijnden, E., Chandler, D.: Solvent coarse-graining and the string method applied to the hydrophobic collapse of a hydrated chain. In: Proceedings of the National Academy of Sciences of the United States of America, vol. 104, pp. 14559–14564 (2007)
65. Santiso, E.E., Trout, B.L.: A general set of order parameters for molecular crystals. J. Chem. Phys. **134**, 064109 (2011)
66. Santiso, E.E., Trout, B.L.: A general method for molecular modeling of nucleation from the melt. J. Chem. Phys. **143**, 174109 (2015)
67. He, X., Shen, Y., Hung, F.R., Santiso, E.E.: Molecular simulation of homogeneous nucleation of crystals of an ionic liquid from the melt. J. Chem. Phys. **143**, 124506 (2015)
68. Barducci, A., Bonomi, M., Parrinello, M.: Metadynamics. Wiley Interdis. Rev. Comput. Mol. Sci. **1**, 826–843 (2011)
69. Allen, F.H.: The Cambridge structural database: a quarter of a million crystal structures and rising. Acta Crystallographica Sect. B-Struct. Sci. **58**, 380–388 (2002)
70. Arduengo, A.J., Dias, H.V.R., Harlow, R.L., Kline, M.: Electronic stabilization of nucleophilic carbenes. J. Am. Chem. Soc. **114**, 5530–5534 (1992)
71. Lopes, J.N.C., Deschamps, J., Padua, A.A.H.: Modeling ionic liquids using a systematic all-atom force field. J. Phys. Chem. B **108**, 2038–2047 (2004)
72. Canongia Lopes, J.N., Padua, A.A.H.: Molecular force field for ionic liquids III: Imidazolium, pyridinium, and phosphonium cations; chloride, bromide, and dicyanamide anions. J. Phys. Chem. B **110**, 19586–19592 (2006)
73. Lopes, J.N.C., Padua, A.A.H.: Molecular force field for ionic liquids composed of triflate or bistriflylimide anions. J. Phys. Chem. B **108**, 16893–16898 (2004)
74. Shimizu, K., Almantariotis, D., Gomes, M.F.C., Padua, A.A.H., Lopes, J.N.C.: Molecular force field for ionic liquids V: hydroxyethylimidazolium, dimethoxy-2-methylimidazolium, and fluoroalkylimidazolium cations and bis(fluorosulfonyl)amide, perfluoroalkanesulfonylamide, and fluoroalkylfluorophosphate anions. J. Phys. Chem. B **114**, 3592–3600 (2010)
75. Lopes, J.N.C., Padua, A.A.H., Shimizu, K.: Molecular force field for ionic liquids IV: trialkylimidazolium and alkoxycarbonyl-imidazolium cations; alkylsulfonate and alkylsulfate anions. J. Phys. Chem. B **112**, 5039–5046 (2008)

76. Fannin, A.A., Floreani, D.A., King, L.A., Landers, J.S., Piersma, B.J., Stech, D.J., Vaughn, R. L., Wilkes, J.S., Williams, J.L.: Properties of 1,3-dialkylimidazolium chloride aluminum-chloride ionic liquids. 2. phase-transitions, densities, electrical conductivities, and viscosities. J. Phys. Chem. **88**, 2614–2621 (1984)
77. Phillips, J.C., Braun, R., Wang, W., Gumbart, J., Tajkhorshid, E., Villa, E., Chipot, C., Skeel, R.D., Kale, L., Schulten, K.: Scalable molecular dynamics with NAMD. J. Comput. Chem. **26**, 1781–1802 (2005)
78. Darden, T., York, D., Pedersen, L.: Particle mesh ewald—an n.log(n) method for ewald sums in large systems. J. Chem. Phys. **98**, 10089–10092 (1993)
79. Hess, B., Bekker, H., Berendsen, H.J.C., Fraaije, J.: LINCS: A linear constraint solver for molecular simulations. J. Comput. Chem. **18**, 1463–1472 (1997)
80. Vanden-Eijnden, E.: Some recent techniques for free energy calculations. J. Comput. Chem. **30**, 1737–1747 (2009)
81. Pruppacher, H.R.: A new look at homogeneous ice nucleation in supercooled water drops. J. Atmos. Sci. **52**, 1924–1933 (1995)
82. Taborek, P.: Nucleation in emulsified supercooled water. Phys. Rev. B **32**, 5902–5906 (1985)

Influence of the Precursor Composition and Reaction Conditions on Raney-Nickel Catalytic System

Sabine Schweizer, Robin Chaudret, Theodora Spyriouni, John Low and Lalitha Subramanian

Abstract Raney-Nickel is routinely used in the process of selective hydrogenation of benzene and its derivatives. In order to gain a better understanding of this catalytic reaction, we have implemented both atomistic and thermodynamic modeling methods. While modeling at the atomistic level provides essential information about structure, electronic effects and dynamics, thermodynamic modeling provides data on physical properties of the system of interest. First, we investigated the influence of the alloy composition on the Raney-Nickel catalyst structure based on a molecular dynamics (MD) based workflow. Different initial and final NiAl compositions were tested. Our simulations indicate that there is a dependence of the pore size on the NiAl composition and this is more pronounced when some Aluminum remains in the catalyst. Next, the solubility of hydrogen in benzene was calculated with thermodynamic modeling. The effect of temperature, pressure, and concentration of cyclohexane (product) on the solubility of hydrogen in benzene was examined. For a given temperature, our studies provided the optimal pressure necessary to obtain maximum solubility of hydrogen in benzene. Finally, based on the results obtained, we have studied the competitive adsorption and chemisorption of benzene and cyclohexane on Raney-Nickel as a first step towards modeling the catalytic hydrogenation of benzene.

Keywords Catalysis · Molecular dynamics · Thermodynamic modeling · Nanostructure · Alloy

S. Schweizer (✉) · R. Chaudret · T. Spyriouni · L. Subramanian
Scienomics, 16 rue de l'Arcade, 75008 Paris, France
e-mail: sabine.schweizer@scienomics.com

J. Low
Argonne National Laboratory, 9700 South Cass Avenue, Argonne, IL 60439, USA

R.Q. Snurr et al. (eds.), *Foundations of Molecular Modeling and Simulation*,
Molecular Modeling and Simulation, DOI 10.1007/978-981-10-1128-3_8

1 Introduction

Catalytic reactions are ubiquitous and play a key role in numerous industrial processes. New application fields, changing demands, and limitations of existing systems make catalysis a challenging research area. In the last decades, computational methods have proven to be useful for revealing critical insights into catalytic systems. Multiscale modeling is becoming increasingly important in this area, as catalytic processes typically involve considerable length and time scales. While questions on physical properties can be addressed by thermodynamic modeling, modeling at the atomistic level can provide essential information about electronic effects, dynamics, and most importantly structural properties and thus allows to pursue a more rational and molecular driven approach for catalyst design. Moreover, amorphous systems are difficult to characterize based on experiments alone and molecular modeling can efficiently complement experimentation.

Here, we present a computational study on Raney-Nickel [1], which is a nanostructured amorphous catalyst used in many industrial applications. It is routinely used in hydrogenation reactions such as the reduction of benzene to cyclohexane. Raney-Nickel is typically prepared by quenching a molten mixture of a NiAl alloy from which Al is leached out for producing the final catalyst. The initial alloy precursor composition is important because it affects the NiAl phases formed during the quenching process. These phases have different leaching properties influencing the porosity of the catalyst and thus its performance.

For gaining a fundamental understanding of this complex catalytic system, one main aspect is to know the structural characteristics of the catalyst and the factors that influence them. Therefore, molecular dynamics (MD) simulations have been performed to study the influence of the alloy composition on the final structure of the catalyst following a recently established workflow for modeling nanoporous structures [2]. In addition to the alloy composition, the impact of the aluminum content in the active catalyst has been examined.

In order to provide comprehensive insights into the catalytic system, we have also performed thermodynamic modeling studies to investigate the reactant mixture in the context of hydrogenation reactions, which belong to the most important applications of Raney-Nickel. The PC-SAFT [3] equation of state has been employed to calculate the influence of temperature and pressure on the solubility of hydrogen in benzene. The effect of the product (cyclohexane) concentration on the hydrogen solubility in benzene was also investigated. This approach will allow us to predict the optimal reaction conditions for obtaining maximum solubility of hydrogen in benzene.

Finally, in the spirit of a multiscale approach, the modeled structures together with the information of the thermodynamic modeling was used to investigate the competitive adsorption and chemisorption of benzene and cyclohexane on Raney-Nickel as a first step towards modeling the catalytic hydrogenation of benzene. Parts of this work involved fairly demanding molecular dynamics simulations enabled by access to high performance computing (HPC) resources at Argonne National Lab (ANL).

2 Computational Details

MAPS [4] software platform was used for generating structures, preparing the MD simulations and analyzing the trajectories. The pore size analysis was carried out using the program package Zeo++ [5, 6]. For the thermodynamic modeling, the SciTherm module of MAPS was used. The MD simulations were carried out using the software package LAMMPS [7] and the LAMMPS-plugin tools in MAPS. For MD simulations on the precursor and final catalyst structures, we used a NiAl alloy potential developed by Mishin [8] which was obtained from the NIST Interatomic Potentials Repository [9]. For studying adsorption and chemisorption processes on the catalyst structure, the reactive force field (ReaxFF) [10, 11] in LAMMPS was employed. The ReaxFF force field allows the creation and breaking of covalent bonds between atoms during a molecular dynamic simulation using a bond index criterion evolving with the interatomic distance. Such a criterion has been extensively described in previous studies [10]. The force-field parameters of the different atoms are updated during the reaction depending on the bond index allowing a modification of all the interaction terms (bond, angle, dihedral, electrostatic, van der Waals…).

For the thermodynamic calculations with PC-SAFT the pure component parameters used in this work are given in Table 1. Binary interaction parameters k_{ij} were fitted for each pair by using phase equilibrium data found in the literature. For the hydrogen/benzene and hydrogen/cyclohexane pairs, the phase equilibrium data [12, 13] were in the temperature range from 339 to 422 K and pressure up to 70 MPa. For benzene/cyclohexane, the data [14] were at 313 and 343 K, and at sub-atmospheric pressure. The following values were found for each pair: k_{ij}(hydrogen/benzene) = 0.37, k_{ij}(hydrogen/cyclohexane) = 0.46 and k_{ij}(benzene-/cyclohexane) = 0.017. The large values of k_{ij} for hydrogen/benzene and hydrogen/cyclohexane reflect the non-ideality of the system due to the difference in size and polarity.

Based on a 2 × 2 × 2 supercell of the conventional crystal structure of $NiAl_3$ (cell parameters were taken from Ref. [15]), starting structures for the precursor

Table 1 PC-SAFT pure component parameters

Compound	m	σ (Å)	ε/k (K)	References
Hydrogen	0.8285	2.973	12.53	[4]
Benzene	2.4653	3.6478	287.35	[3]
Cyclohexane	2.5303	3.8499	278.11	[3]

models have been generated by randomly replacing Ni and Al, respectively, in the corresponding fractions. In total four different initial NiAl compositions were considered: (1) 40 wt% Ni—60 wt% Al (denoted as Ni30Al98), (2) 50 wt% Ni—50 wt% Al (denoted as Ni40Al88, results are discussed in Ref. [2]), (3) 60 wt % Ni—40 wt% Al (denoted as Ni52Al76), (4) 70 wt% Ni—30 wt% Al (denoted as Ni66Al62). For each one of the compositions, five different models were created. These models have been enlarged to 8 × 8 × 8 supercells each containing more than 65,000 atoms with cell lengths between 70 and 120 Å. The structures were then subjected to several MD simulations to generate porous structures. First, the structures were equilibrated over 50 ps using an NVT ensemble. Subsequently, a 200 ps equilibration using an NPT ensemble at 2000 K was performed, before quenching the system to room temperature over 1 ns. To confirm the convergence of the simulations after quenching, an additional 1 ns equilibration was performed using an NPT ensemble at 300 K for one configuration per composition. Since for each composition similar results were obtained for all five models, subsequent calculations were performed for only one structure per composition. Aluminum was removed to make three different fractions: (a) 0 %, (b) 5 % Al, and (c) 10 % Al. After a short geometry optimization, equilibrations over 10 ns with NVT ensemble and 15 ns with NPT ensemble were carried out at 300 K. A more detailed description of the workflow can be found in Ref. [2].

The final Raney-Nickel system from the industrially widely used 50 wt% Ni—50 wt% Al precursor with 0 % of remaining Al was used as catalyst for the ReaxFF simulation. As a preliminary study of the catalytic reaction, we focused on modeling the adsorption of benzene on Nickel surface. For this purpose, MAPS platform was used to set up an initial system of 100 benzene molecules within the porous catalyst. The system was heated to 500 K and simulated at the same temperature for 25 ps. For comparison, Nickel (111) and Nickel (100) surfaces were built. The Nickel (111) surface contains 6 layers of 8 × 8 Nickel atoms for a total 384 atoms on top of which 20 Å of vacuum were added. The simulation unit had the following dimension $a = b = 20$ Å and $c = 30$ Å with angles $\alpha = \beta = 90°$ and $\gamma = 120°$. The Nickel (100) surface contains 9 layers of 8 × 8 Nickel atoms for a total of 288 atoms on top of which 20 Å of vacuum were added. The box had the following dimension $a = b = 20$ Å and $c = 30$ Å with angles $\alpha = \beta = 90°$ and $\gamma = 120°$. The vacuum was filled with 45 and 35 benzene molecule for Nickel (111) and Nickel (100) surface, respectively. The analysis of the ReaxFF simulation was performed using ReaxFF plugin module within MAPS.

3 Results and Discussion

In the first sub-section we present the results obtained for the different catalyst compositions. In the second sub-section, the thermodynamic modeling studies will be shown, and in the last part results obtained on the reactivity are discussed.

3.1 *Molecular Dynamics Simulation on Different Catalyst Compositions*

One main goal of the present work was to investigate how the precursor composition affects the structure of the active catalyst.

In the actual catalyst, different NiAl phases are present. We started from a simplified model and have used only one phase as a building block to keep the system size and computational effort reasonable. More advanced models including different phases will be subject of future work and are beyond the scope of the present work.

After quenching to room temperature, the density and cell parameters of the different alloy models were compared for each composition. Depending on the composition, the density values of the precursor range between 3.9 g/cm^3 and 5.9 g/cm^3 which is in line with experimentally determined values of 3.95 g/cm^3 and 4.76 g/cm^3 for $NiAl_3$ and Ni_2Al_3 [16]. For the different initial configurations of each composition we found the same values in each case. The respective densities remained constant after an additional 1 ns equilibration step using NPT ensemble confirming the convergence of the simulations. The cell parameters also behaved consistently with a standard deviation of 9 Å from the average value at most.

After quenching the alloy structures to 300 K, aluminum was removed in three different fractions to emulate the experimental leaching process: (a) None of the aluminum was kept (denoted as 0 % Al). (b) 5 % aluminum was kept (denoted as 5 % Al), and (c) 10 % aluminum was retained (denoted as 10 % Al) in order to investigate the influence of the remaining aluminum on the structural characteristic of the final catalyst. After removing the aluminum, the structures were geometry optimized and then subjected to a MD simulation using NVT ensemble over 10 ns at 300 K to allow for pore formation. For relaxing the cell volume, a 15 ns MD simulation using NPT ensemble was performed at the same temperature. In Ref. [2], the structural properties of the 0 and 5 % Al models based on the initial 50 wt% Ni model were thoroughly characterized and compared with experimental results. The computed properties show a good agreement with experimental data validating the computational approach for modeling the catalyst structure. In the following, we will now compare the structural characteristics for different initial and final compositions. The structures were analyzed with respect to the final densities and the pore sizes. The final densities are listed in Table 2. As expected the density increases with increasing amount of Ni with respect to the initial composition. Interestingly, for a given composition, the density decreases with increasing amount

Table 2 Densities in g/cm^3 after 15 ns NPT MD simulation

Remaining Al (%)	Initial wt% Al			
	60	50	40	30
0	4.2	4.1	5.4	5.9
5	2.1	3.9	5.2	5.7
10	2.6	3.8	4.1	5.0

of remaining aluminum which could indicate a stabilizing influence of Al on the overall structure. This trend is observed for each composition except for the 60 wt% Al composition for which a slightly smaller density is found for 5 % Al structure than for the 10 % Al structure.

The results of the pore size analysis are summarized in Fig. 1 in which the largest pore size diameter after NVT and NPT equilibration are illustrated. The removal of the Al atoms causes a fairly large void volume. In order to allow a decent structural equilibration, first the volume of the simulation box was kept constant which induced the pore formation. In the NPT simulation performed subsequently, the volume was allowed to equilibrate under constant pressure which is typically closer to experimental conditions in this context. As can be expected, the volume reduced during the NPT equilibration, but the simulation box did not collapse. Consequently, the largest pore size diameter reduced during the NPT run (orange vs. red /purple vs. green /cyan vs. blue bar in Fig. 1). For each initial Ni/Al composition (60/50/40/30 wt% Al) we have studied three Al fractions (0, 5, 10 %) for the final catalyst. If we compare the results after NPT simulation for the three Al fractions, we observe that the largest pore size diameters are obtained for the 5 % Al structures (green vs. red and blue bars in Fig. 1), except for the initial 30 wt% Al composition. For this composition, the largest pore size diameter was found for the 10 % Al structure (blue bar in Fig. 1), while the maximum pore size diameters of the 0 and 5 % model are almost similar (red and green bar in Fig. 1). This finding further confirms the assumption that aluminum has a stabilizing influence on the porosity of the final catalyst. The largest pore size diameter of the 0 % Al structures (red bars in Fig. 1) ranges between 34 and 44 Å, of the 5 % Al structures (green bars in Fig. 1) between 32 and 69 Å, and of the 10 % Al structures (blue bars in Fig. 1) between 43 and 58 Å. For the 0 % Al structures (red bars in Fig. 1), the largest pore size was found for the initial 50 wt% Al composition, while for the 5 and 10 % Al structure (green and blue bars in Fig. 1) the largest pore size diameter was obtained for the initial 60 wt% Al composition. The 10 % Al structures (blue bars in Fig. 1) shows an increasing largest pore size diameter from the initial 50 wt% Al composition to the 30 wt%

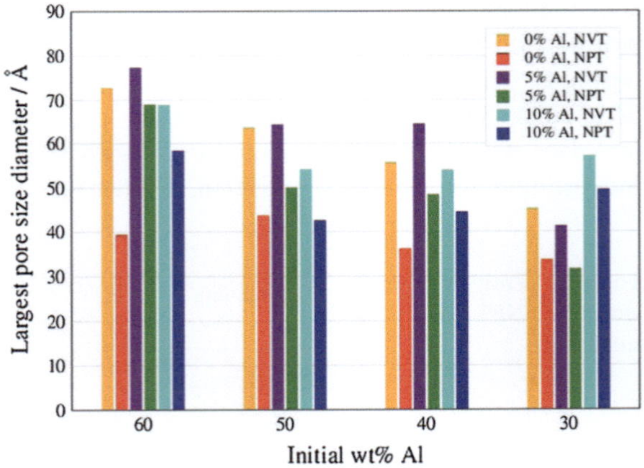

Fig. 1 Largest pore size diameter in Å after 10 ns NVT and 15 ns NPT equilibration

Al composition, while for theses compositions a decrease in the largest pore size with decreasing Al content was observed for the 0 and 5 % Al structures (red and green bars in Fig. 1). Zeifert el al. [17] compared alternative experimental routes for the catalyst preparation and determined the mean pore size diameter for different initial NiAl compositions. For initial alloy compositions containing 58/50/46 wt% Al, the mean pore diameters were obtained as 37/55/27 Å. Experimentally, there is thus also no uniform dependence observed of the pore size on the initial Al content. Candy and Fouilloux [18] measured a pore size diameter of 45 Å for a Raney Ni sample containing 4.7 % Al. Basically, our results are in reasonable agreement with experimental data bearing in mind that the mean pore diameter is certainly smaller than the largest pore size diameter used in our analysis. Our results suggest that there is a complex balance between the maximum pore size and the aluminum content in the initial and final structure, respectively.

Overall, the analysis of the MD simulations clearly suggests that the porosity in terms of the pore size diameter depends on the initial composition and on the remaining aluminum content after activation of the catalyst.

3.2 *Thermodynamic Modeling of the Reactants*

In addition to the simulation of the catalyst, we also performed thermodynamic modeling studies on the reactants and products in order to investigate physical properties during the catalytic reaction. The solubility of hydrogen in benzene and benzene/cyclohexane mixtures was calculated with PC-SAFT at the temperature range 250–523 K. T-P Flash calculations were performed at constant pressure 10, 20 and 50 bar. The results are illustrated in Fig. 2. The composition of the liquid phase in these T-P Flash calculations is not kept constant, therefore the mole fraction of cyclohexane shown in Fig. 2 (0.36 (top) and 0.96 (bottom)) is approximate.

As expected, the solubility of hydrogen increases with increasing pressure. The solubility shows maxima with temperature at the various pressures. The highest solubility is observed at 475 K and 50 bar. Finally, the presence of cyclohexane results in an increased hydrogen solubility in all cases.

3.3 *Reactive Force Field Simulations on Benzene Adsorption on the Raney-Nickel Catalyst*

Based on the results presented above, we have performed preliminary simulations on the catalytic reaction. For this purpose, we have studied the adsorption of benzene on Raney-Nickel using one of the model structures generated as discussed

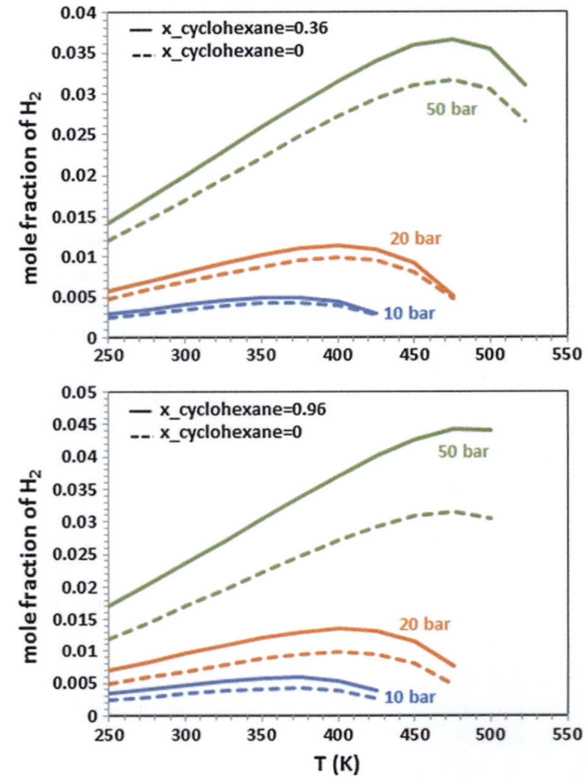

Fig. 2 Results of PC-SAFT for hydrogen solubility (mole fraction) in benzene (*dashed lines*) and benzene/cyclohexane mixtures (*solid lines*) in the temperature range 250–550 K and at 10, 20, and 50 bar. The mole fraction of cyclohexane is approximately 0.36 (*top*) and 0.96 (*bottom*)

above and compared the results for Raney-Nickel with results obtained for Ni (100) surface and Ni (111) surface. A snapshot showing benzene adsorption in Raney Ni catalyst is illustrated in Fig. 3.

The results presented in Fig. 4 show the evolution of the number of adsorbed benzene in the different system as a function of the simulation time. The ReaxFF force field allows the creation and breaking of covalent bonds between the different atoms of the system during the molecular dynamics simulation. In this work, we considered that a benzene molecule was adsorbed, if it formed at least one bond with the Ni (100), Ni (111), or Raney-Nickel surface, respectively. In the first 5 ps, the benzene adsorption is comparable for all three systems evaluated, i.e. both clean Ni surfaces and Raney Ni model. After the first 5 ps, about 6–7 % of benzene molecules have been adsorbed. In the very beginning of the simulation time, the adsorption process is even faster on Ni (100) and Ni (111) surface (blue and green line in Fig. 4) compared to the catalyst (purple line). After the first few ps, the adsorption on Ni (100) surface (blue line in Fig. 4) remains rather constant and does not increase much. After 25 ps, only about 12 % of benzene have been adsorbed. In contrast to this finding, the benzene adsorption on the Ni (111) surface and the Raney Ni model system (green and purple line in Fig. 4) increases more

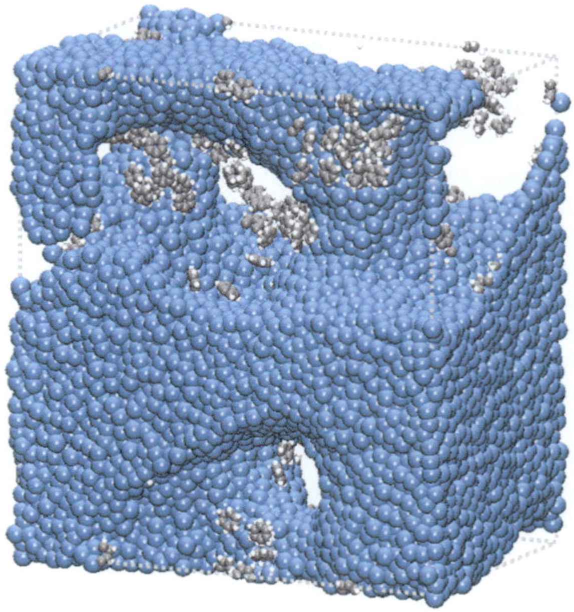

Fig. 3 Snapshot of a ReaxFF simulation of benzene adsorption in Raney Ni. The model structure of Raney Ni was generated as described above for an initial 50 wt% Ni/Al composition

continuously after the first few ps of the simulation, while the increase is steeper for the Raney Ni system (purple line in Fig. 4). After 25 ps adsorption of benzene in Raney-Nickel appears thus faster than on both Nickel surfaces with more than 20 % of benzene molecules adsorbed.

The analysis of the reaction profiles indicates that the adsorption is not finished and longer simulation times are necessary to model the catalytic reaction. These early results are however promising for modeling benzene adsorption and

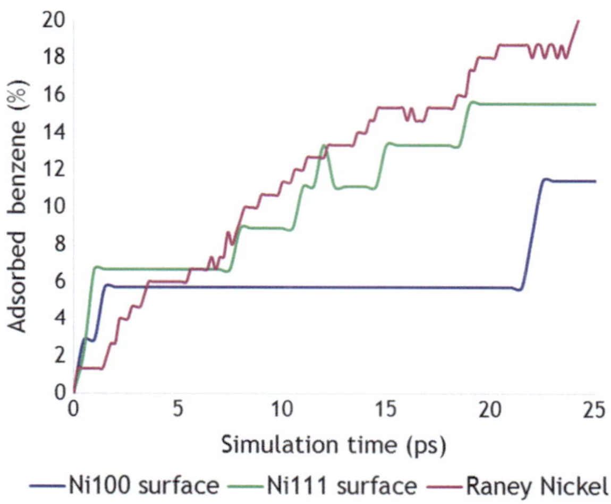

Fig. 4 Adsorbed benzene in percent over simulation time for Ni (100) surface, Ni (111) surface, and Raney Ni

hydrogenation reaction kinetic on Raney-Nickel. Jobic et al. [19] have reported experimental studies on benzene-nickel systems using neutron inelastic spectroscopy and found a stronger bonding of benzene on Raney Ni compared to that on Ni (111) and Ni (100). This finding matches well with the computational results indicating that more benzene is adsorbed on Raney Ni assuming that a higher coverage can be expected when benzene is bonded stronger. Experimentally [19], a higher force constant was found for Ni (100) compared to Ni (111), while the simulations indicate a slightly higher coverage on Ni (111). However, this may be attributed to the short simulations times. Future studies will include longer simulations of pure benzene and hydrogen/benzene mixture will be studied using ReaxFF. For these simulations, it has become apparent during our preliminary studies that the ReaxFF force field parameter set needs to be further refined, which is, however, beyond the scope of the present work. In subsequent work, we plan to refine the parameters and also utilize the results from thermodynamic modeling work for setting up the system and the simulation conditions. In addition to the adsorption behavior, the influence of the Al content on the adsorption and reaction kinetics will then be investigated as well.

4 Conclusions

In this work, we have presented a molecular modeling study on Raney-Nickel. Our results show a dependence of the pore size on the initial precursor and final catalyst compositions. The simulations indicate a stabilizing influence of the aluminum on the remaining porosity. In addition, thermodynamic modeling of physical properties of a possible reactant mixture provide insights into optimal initial reaction conditions. Furthermore, we have shown results on the chemisorption of benzene on Raney-Nickel which were compared to the adsorption on a conventional clean Ni surface.

Acknowledgments The authors would like to thank Adri van Duin for his help with the ReaxFF potential. We gratefully acknowledge the computing resources provided on Blues and Fusion, high-performance computing cluster operated by the Laboratory Computing Resource Center at Argonne National Laboratory. L.S. would like to thank Raymond Bair at Argonne National Lab for providing HPC resources for this work.

References

1. Raney, M.: US Patent 1,628,190 A (1927)
2. Schweizer, S., Chaudret, R., Low, J., Subramanian, L.: Molecular modeling and simulation of Raney Nickel: from alloy precursor to the final porous catalyst. Comp. Mater. Sci. **99**, 336–342 (2015)

3. Gross, J., Sadowski, G.: Perturbed-chain SAFT: an equation of state based on a perturbation theory for chain molecules. Ind. Eng. Chem. Res. **40**, 1244–1260 (2001)
4. Scienomics, MAPS Platform: Version 3.4.1, France, (2014)
5. Martin, R.L., Smit, B., Haranczyk, M.: Addressing challenges of identifying geometrically diverse sets of crystalline porous materials. J. Chem. Inf. Model. **52**, 308–318 (2012)
6. Willems, T.F., Rycroft, C.H., Kazi, M., Meza, J.C., Haranczyk, M.: Algorithms and tools for high-throughput geometry-based analysis of crystalline porous materials. Microporous Mesoporous Mater. **149**, 134–141 (2012)
7. Plimpton, S.: Fast parallel algorithms for short-range molecular dynamics. J. Comput. Phys. **117**, 1–19 (1995)
8. Mishin, Y.: Atomistic modeling of the γ and γ'-phases of the Ni–Al system. Acta Mater. **52**, 1451–1467 (2004)
9. Becker, C.A., Tavazza, F., Trautt, Z.T., Buarque de Macedo, R.A.: Considerations for choosing and using force fields and interatomic potentials in materials science and engineering. Curr. Opin. Solid State Mater. Sci. **17**, 277–283 (2013)
10. van Duin, A.C.T., Dasgupta, S., Lorant, F., Goddard III, W.A.: ReaxFF: a reactive force field for hydrocarbons. J. Phys. Chem. A **105**, 9396–9409 (2001)
11. Mueller, J.E., van Duin, A.C.T., Goddard, W.A.: Application of the ReaxFF reactive force field to reactive dynamics of hydrocarbon chemisorption and decomposition. J. Phys. Chem. C **114**, 5675–5685 (2010)
12. Thompson, R.E., Edmister, W.C.: Vapor-liquid equilibria in hydrogen-benzene and hydrogen-cyclohexane mixtures. A.I.Ch.E. J. **11**, 457–461 (1965)
13. Brainard, A.J., Williams, G.B.: Vapor-liquid equilibrium for the system hydrogen-benzene-cyclohexane-n-hexane. A.I.Ch.E. J. **13**, 60–69 (1967)
14. Scatchard, G., Wood, S.E., Mochel, J.M.: Vapor-liquid equilibrium III. Benzene-cyclohexane mixtures. J. Phys. Chem. **43**, 119–130 (1939)
15. Khare, N.P., Lucas, B., Seavey, K.C., Liu, Y.A.: Steady-state and dynamic modeling of gas-phase polypropylene processes using stirred-bed reactors. Ind. Eng. Chem. Res. **43**, 884–900 (2004)
16. Bradley, A.J., Taylor, A.: The crystal structures of Ni_2Al_3 and $NiAl_3$. Philos. Mag. Series **7**, 1049–1067 (1937)
17. Zeifert, B., Blásquez, J.S., Moreno, J.G.C., Calderón, H.A.: Raney-Nickel catalysts produced by mechanical alloying. Rev. Adv. Mater. Sci. **18**, 632–638 (2008)
18. Candy, J.P., Fouilloux, P.: Adsorption and hydrogenation of benzene-^{14}C vapor on Raney Nickel. J. Cat. **38**, 110–119 (1975)
19. Jobic, H., Tomkinson, J., Candy, J.P., Fouilloux, P., Renouprez, A.J.: The structure of benzene chemisorbed on Raney Nickel: a neutron inelastic spectroscopy determination. Surface Sci. **95**, 496–510 (1980)

Atomistic Modeling and Simulation for Solving Gas Extraction Problems

Genri E. Norman, Vasily V. Pisarev, Grigory S. Smirnov and Vladimir V. Stegailov

Abstract Proof-of-concept results are presented on the application of molecular modeling and simulation to the gas extraction problems. Both hydrocarbon mixtures and gas hydrates in porous media are considered. Retrograde gas condensation reduces the amount of recoverable gas in reservoirs and can lead to jamming of wells. For example, the authors [1] developed a model of two-phase gas filtration through porous media that can reproduce the jamming. The model can describe gas flow in soil of reservoir if both a phase diagram of the gas mixture and permeability of pores to gaseous and liquid phases are known. Molecular dynamics simulations are used to study phase diagrams of binary hydrocarbon mixtures at temperatures between the critical points of pure components. The phase diagrams in free space and in slit pores are calculated. Effects of wall–gas interaction on the phase diagram are estimated. The data obtained from molecular simulations can be used to improve the hydrodynamic filtration model and to optimize the natural gas and gas condensate extraction conditions. Effects of pore structure on the phase stability of gas hydrates and on the diffusion of guest molecules are studied by means of molecular modeling. The anisotropic diffusion is found in hydrogen hydrates. Moreover, diffusivity of hydrogen molecules demonstrates anomalous behavior on nanosecond timescale.

Keywords Phase diagrams · Methane · Molecular dynamics · Clathrate hydrates · Retrograde condensation · Porosity

G.E. Norman · V.V. Pisarev (✉) · G.S. Smirnov · V.V. Stegailov
Joint Institute for High Temperatures of RAS, 125412 Moscow, Russia
e-mail: pisarevvv@gmail.com

G.S. Smirnov
Moscow Institute of Physics and Technology (State University),
141700 Dolgoprudnyy, Moscow Region, Russia

R.Q. Snurr et al. (eds.), *Foundations of Molecular Modeling and Simulation*,
Molecular Modeling and Simulation, DOI 10.1007/978-981-10-1128-3_9

1 Introduction

Natural gas extraction and storage give rise to a number of scientific and technical problems which require the knowledge of gas mixture behavior in porous media. Two particular problems are the modeling of hydrocarbon filtration through porous reservoir rocks and the modeling of gas–water systems at high pressures. They require the knowledge of phase diagrams and transport coefficients of multicomponent systems. Phase diagrams of one-component substances on the pressure (P)–temperature (T) plane consist of two-phase coexistence lines. Single-phase stability areas span between the lines. In contrast, a two-phase coexistence region transforms into a surface in the (P-T-α) space for binary mixtures phase diagrams, where α is the molar fraction of one of the components. In particular, there is a region of the two-phase surface for binary and multicomponent mixtures, where gas phase partially condenses into a liquid at the isothermal depressurizing. This phenomenon is known as retrograde condensation, and it can occur at temperatures higher than a critical temperature of the most volatile component of a mixture.

Since the critical temperature of methane is 190.6 K, methane-containing mixtures at ambient temperatures always have some range of methane concentrations at which the retrograde condensation occurs. This condensation complicates gas field operation, since it results in partial condensation of natural gas near a well bottom. It lowers a well yield and decreases the amount of recoverable hydrocarbons.

Gas filtration through a porous medium is often described mathematically in the form of the Darcy equation $u = KI$, where u is a filtration rate, I is a head gradient, and permeability coefficient K is the main characteristics of the medium. To model gas reservoirs, it is necessary to know permeability coefficients for both gas and liquid phases and to have a model to calculate reservoir liquid saturation [1, 2]. The equilibrium liquid saturation depends only on the thermodynamic functions of the fluids and reservoir walls.

The bulk phase diagrams of pure hydrocarbons and mixtures are well known from the experiments. In the work by Sage et al. [3], the bubble point pressures of methane + n-butane mixtures are determined experimentally from the discontinuity of isothermal compressibility of constant-composition mixture at the point of phase transition. The composition of vapor phase is determined in that work from the residual specific volume of gas. Later experiments employ phase recirculation techniques [4] to achieve vapor–liquid equilibrium [5, 6], and the phase compositions are analyzed by more advanced methods such as gas chromatography.

Molecule–wall interaction may shift phase diagrams in porous media, especially in nanopores. One of the ways to quantify such changes is the calculation of phase diagrams via atomistic simulations using molecular dynamics (MD) [7, 8] or Monte Carlo (MC) [9–13] methods. Semigrand ensemble simulations [12] and Gibbs–Duhem integration [13] are the most used MC techniques to tackle with the problem of multicomponent mixture phase diagram calculation. The MC approach is good for purely thermodynamic properties, but it does not allow calculation of dynamic properties. MD method is widely used for the calculation of phase diagrams [14], structure [15–18], and transport properties [19, 20] of confined fluids. In the present work, we use the MD method for phase diagram calculations to validate the potential model. We plan to use the same potential model then in the future works to calculate the transport properties.

The phase diagram of a model methane + n-butane hydrocarbon mixture is studied in this paper. Such components are chosen because this mixture reproduces qualitatively phase diagram peculiarities of more complex hydrocarbons with the retrograde condensation, on one hand, and is studied experimentally at the Plast setup [1, 2] in the Joint Institute for High Temperatures of RAS, on the other hand. Due to the large critical temperature difference (190.6 K for methane vs. 425.1 K for butane), vapors below the critical point of methane have vanishing butane concentrations. Because of that, molecular simulations of phase equilibrium would require impractically large number of particles. For the gas extraction tasks, the supercritical region with respect to methane poses the greatest interest.

Modeling of natural gas with high water content poses an additional problem of gas hydrate formation. Clathrate gas hydrates are crystalline water-based inclusion compounds physically resembling ice. They require elevated pressures and low temperatures to be formed and are found in gas pipelines, permafrost regions, ocean sediments, comets, and certain outer planets [21, 22]. Guest molecules are trapped inside cavities, or cages, of the hydrogen-bonded water framework. The clathrate structure type is mainly determined by the size of guest molecules. Gas hydrates allow compact storage of hydrocarbons since one volume of hydrate may contain 180 volumes of gas (STP). The discovery of hydrogen hydrates (HH) attracted significant attention to the H_2 + H_2O phase diagram and clathrate structures. Along with the fundamental interest and significance for geophysics of icy moons and outer planets, HH provide a way to prospective hydrogen storage technologies. Diffusion of guest molecules plays a key role at hydrate storage and transportation. It affects the saturation of crystals with surrounding gases as well as the kinetics of clathrate decay and formation.

The diffusivity of hydrogen molecules is mostly studied for hexagonal ice and sII clathrate structure. Strauss [23] showed by neutron inelastic scattering that the diffusion coefficient of H_2 in deuterated ice at 25–60 K is rather high and comparable with the self-diffusion coefficient in liquid hydrogen. About 272 K, hydrogen solubility in hexagonal ice is comparable with that in liquid water at atmospheric pressure and differs by two times at 100 MPa [24].

In the clathrate structure sII, the diffusion coefficient of hydrogen molecules is lower by several orders of magnitude. The modeling of process of hydrogen molecule diffusion is performed previously in works [25–29]. Gorman et al. [28] reveal two diffusion modes in the sII structure: diffusion within one cavity on short timescale and "jumps" between cavities on long timescale.

The importance of hydrates requires the accurate knowledge of their thermodynamic and kinetic properties, mechanisms of formation and decay. Molecular simulation is a method of choice for such theoretical studies since it can explicitly capture the structure of gas hydrates and their constituents. MD is used to study different processes in methane hydrate, especially phase diagram of hydrates. Tung et al. [30] determined the coexistence line in a wide range of pressures using TIP4P/Ew water model and OPLS-AA model for methane. They analyzed the evolution of the potential energy as a function of time during NPT simulations. Conde and Vega [31] used a similar technique to determine the coexistence points at up to 400 bar. They established that the TIP4P/Ice model [32] gives the best agreement with the experimental results, but their results differ from the Monte Carlo data of Jensen et al. [33]. In our previous work [34], we confirm the data of Jensen et al. [33] and suggest that TIP4P/2005 [35] water model gives better results at higher pressures than TIP4P/Ice.

Section 2 is devoted to the simulation details used at modeling of both problems. Simulation results are presented in Sect. 3. Phase diagrams for both bulk and porous systems are treated for gas condensates in Sects. 3.1 and 3.2. Melting and decay of the superheated sI methane hydrate structure are studied using MD simulation in Sect. 3.3. The melting curve is calculated by the direct coexistence simulations in a wide range of pressures up to 5 kbar for the SPC/E, TIP4P/2005, and TIP4P/Ice water models and the united atom model for methane. We also discuss diffusion of guest molecules in hydrogen hydrates in Sect. 3.4.

2 Simulation Details

2.1 Methane + n-Butane Mixture

TraPPE-UA (Transferrable Potential for Phase Equilibria–United Atom model) force field [36] is used for methane + n-butane mixtures. Methane molecules are presented by point particles, and butane molecules are reduced to four-particle models. Due to complications with rigid bonds in the MD method, a fully flexible butane molecule model is used instead of rigid bonds suggested in the original TraPPE force field. Spring constants for C–C bonds are taken from the AMBER force field [37]. Force field authors claim that phase diagrams are determined mainly by the intermolecular forces so such augmentation would still give the correct phase equilibrium [38]. Lennard-Jones (LJ) potential is used for nonbonded interactions. The LJ cutoff radius is 16 Å, and the potential and its derivative are smoothed to zero from 16 to 18 Å.

rRESPA scheme [39] is used for the numerical integration of motion equations. 4 fs timestep is chosen for nonbonded interactions, 2 fs for dihedral torsions, and 1 fs for bond and angle oscillations. Periodic boundary conditions are employed. The MD box size is chosen as $15 \times 15 \times N a_0^3$, where $a_0 = 6.8$ Å is the parameter of a simple cubic lattice and $N = 80$–250 depends on the target pressure and fraction of methane.

A following approach is applied to create a two-phase gas–liquid system. First, 9000 butane molecules are placed in the simple cubic lattice sites of the volume $15 \times 15 \times 40 a_0^3$. The z-axis is a preferential direction in this configuration, which is normal to the interface. The whole box volume is filled then with randomly distributed methane molecules. Energy minimization is then applied to relax the structure and move apart the particles which are generated unphysically close to each other. The number of methane molecules defines the mixture composition. Mixtures with 25–70 molar percentage of methane are considered.

Nose–Hoover thermostat [40] and Shinoda barostat [41] are applied at MD runs. As a fluid medium is simulated, the external pressure is established by changing only the L_z size of the simulation box to fit the P_{zz} pressure tensor component to the target value. The sizes L_x and L_y remain the same during the simulation, and the isotropic stress tensor is maintained hydrostatically by the fluid phases. The simulations are carried out for 1.5 million timesteps, or 6 ns, and component densities are then averaged over the last 500,000 timesteps in 100 bins along the z-axis to obtain the profiles.

2.2 *Gas Hydrates*

Although even simplified potential models can capture some important features of water, we have to use state-of-the-art classical potentials for accurate overall description of the water phase diagram in the solid phase. We use the TIP4P/Ice [32] model that gives a very good description of the ice phase coexistence lines. We consider for comparison TIP4P/2005 [35] and a well-known SPC/E [42] model. SPC/E is a simple 3-site model with charges located on H and O sites. TIP4P models are the 4-site models with a negative massless charge located near the oxygen atom and positive charges located on H atoms. We use a simple LJ model for methane and three-site model with charges for hydrogen molecules. The cross-interaction between guest and host molecules is described by the Lorentz–Berthelot rules:

$$\varepsilon_{ij} = \chi\sqrt{\varepsilon_{ii}\varepsilon_{jj}}, \sigma_{ij} = \left(\sigma_{ii} + \sigma_{jj}\right)/2, \qquad (1)$$

where ε_{ii} and σ_{ii} are the LJ parameters for the pure ith component, and ε_{ij} and σ_{ij} are the cross-interaction parameters, where $\chi = 1$ or $\chi = 1.07$. The latter value indirectly introduces polarization of methane in TIP4P/2005 water [43].

We use 9 Å cutoff distance for the LJ interactions. The PPPM algorithm is applied to take into account the long-range interactions, with 9 Å cutoff for the real-space part. Water molecule bonds and angles are fixed using the SHAKE algorithm. The 3D periodic boundary conditions are used. The integration time step is 2 fs.

Diffusion of hydrogen molecules in hydrates is studied using the classical MD method. Our previous work shows the stability of C_0 and sT′ structures at pressures 2–10 atm. We study in the current work diffusion of hydrogen molecules at the pressure 0.6 GPa and temperatures from 140 to 260 K. The previous works show that both structures are stable at those conditions at 100 % cage occupancy regardless of the potential used.

The simulations are performed with the Nose–Hoover thermostat and Shinoda barostat. The MSDs are calculated by both time averaging along the individual MD trajectories and ensemble averaging over several trajectories.

We determine first the equilibrium temperatures and pressures for coexistence. Conde and Vega in their work [31] performed similar calculations using long NPT MD trajectories (up to 1 μs). They waited for complete crystallization or complete melting of the initial three-phase system at several fixed temperatures. We follow another approach looking directly for the phase coexistence conditions.

We start from a $5 \times 5 \times 10$ unit cells clathrate system. Then, we keep atoms in one half of the cell frozen on their positions and raise the temperature in the other half to melt it. After such a procedure, we have the initial nonequilibrium system. Then, a short NPT simulation is performed to drive the system to the desired temperature and pressure. Finally, we perform a several nanosecond-long NVE MD simulation. The sI phase grows or melts in this simulation depending on whether we overshot or undershot the clathrate melting temperature at the given pressure. After partial melting or crystallizing, the system should stabilize at some temperature and pressure corresponding to the equilibrium curve. Reaching the equilibrium during crystallization requires longer simulation times, especially when methane molecules form a bubble in water. In our calculations, we consider only the cases when clathrate melts and equilibrium establishes much faster, during a few nanoseconds. When the volume of the sI phase stops changing, we assume that the temperature and pressure in the system correspond to the coexistence conditions.

All MD simulations are conducted using the LAMMPS package [44].

3 Simulation Results

3.1 Gas Condensates: Bulk Simulations

Density profiles of the hydrocarbon mixture components are calculated by MD simulations. The examples of the profiles are presented in Fig. 1 for temperature 330 K at two pressures. The liquid phase is butane-rich; the vapor phase is methane-rich. The density profiles turn out to be non-typical for a liquid film. Absolute value of the methane density does not change remarkably at the transition from vapor to liquid phase. Moreover, the absolute methane density is lower in liquid phase with respect to vapor at some conditions. It is interesting to note that there is a maximum of the methane density near the phase boundary at 40 atm. It points to the methane adsorption on the interface. Similar phenomena are observed at the modeling of the liquid in a contact with solid walls [15–17, 20], as well as in Coulomb clusters [45, 46].

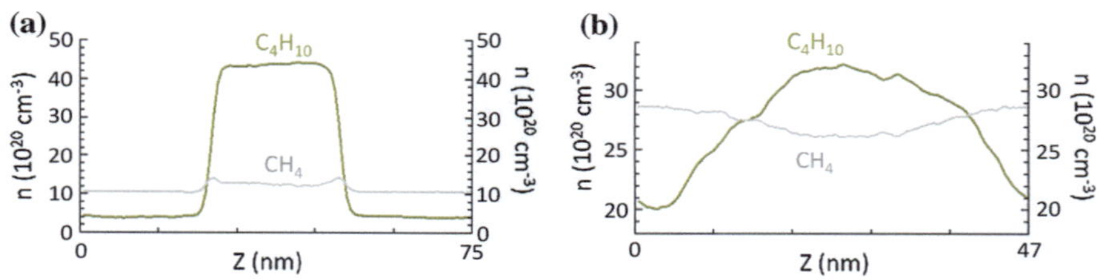

Fig. 1 Component density profiles in vapor–liquid coexistence simulations for methane + n-butane mixture at 330 K: 40 atm (**a**) and 90 atm (**b**)

Fig. 2 Phase equilibrium curve of methane + n-butane at 330 K: MD model compared to the experimental data [3]. The error bars show statistical uncertainties. The error bars lie within the symbols if not shown

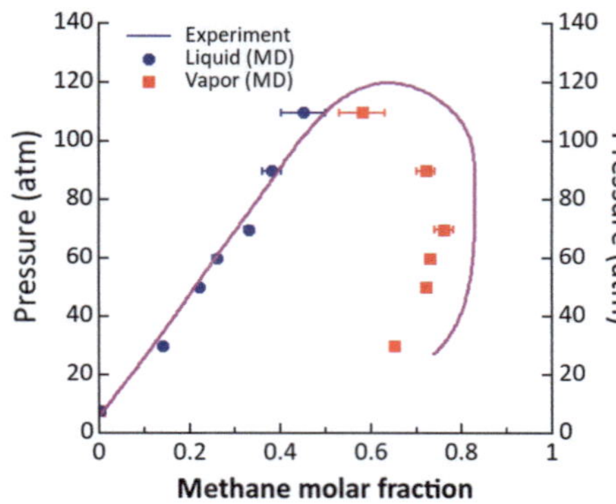

The phase equilibrium curve is calculated for the methane + n-butane mixture at 330 K (Fig. 2). The force field model used reproduces experimental data [3] on methane solubility in liquid butane rather well up to 80 atm. It reproduces the existence of the retrograde condensation region for the mixture under consideration at this temperature. The existence of the region follows from the fact that the phase equilibrium curve does not reach 100 % methane molar fraction.

3.2 *Gas Condensates: Pore Simulations*

Calculations are also performed for the phase equilibrium in a pore. The simplest model of a slit pore with smooth walls is considered. Interaction of a particle with walls is taken in the Lennard-Jones 9-3 form

$$U_{\text{wall}}(r) = \varepsilon_W \left[\frac{2}{15} \left(\frac{\sigma_W}{r} \right)^9 - \left(\frac{\sigma_W}{r} \right)^3 \right],$$

where r is a distance from a particle to the wall, and σ_w and ε_w are potential parameters. The walls are perpendicular to the x-axis. The examples of wall surfaces with ε_w = 0.35 kcal/mol, σ_w = 0.35 nm ("weak" wall potential) and ε_w = 0.5 kcal/mol, σ_w =

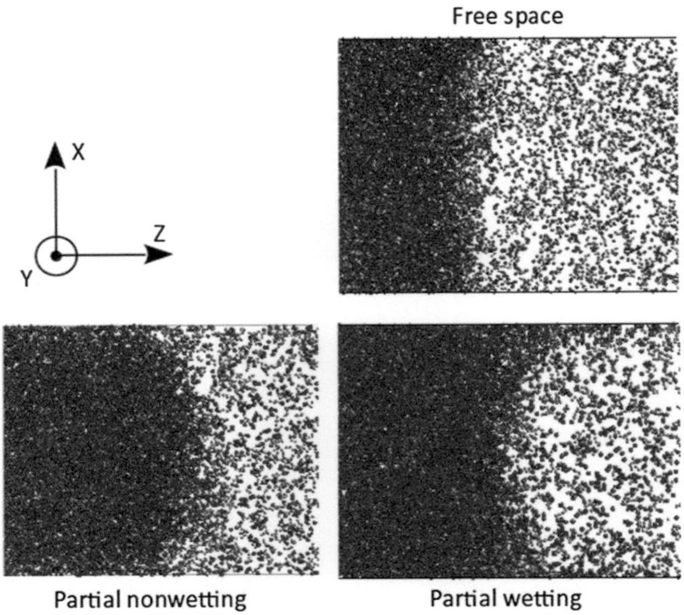

Fig. 3 Shape of the interphase boundary in the *XZ* plane for walls with different wettabilities at 330 K and 30 atm. Butane molecules are shown in *green* and methane molecules in *blue*. Every picture shows two overlaid snapshots of the simulation cell

0.39 nm ("strong" wall potential) are considered. The LJ parameters for the mixture components are $\varepsilon_{CH4} = 0.29$ kcal/mol and $\sigma_{CH4} = 0.373$ nm, $\varepsilon_{CH3} = 0.19$ kcal/mol and $\sigma_{CH3} = 0.375$ nm, and $\varepsilon_{CH2} = 0.09$ kcal/mol and $\sigma_{CH2} = 0.395$ nm. The simulations are conducted with the distances 4.08 and 10.2 nm between walls.

To obtain the density profiles in pores, we used longer simulations, for 2.5 million timesteps, or 10 ns. Density is averaged over the last 200,000 timesteps in 100 bins to obtain the profiles.

In the 10.2 nm pores, the different wettability of the walls is clearly seen. "Weak" walls show contact angle >90°, and "strong" walls show contact angle <90° (Fig. 3), which means partial wettability of the "strong" walls and partial nonwettability of "weak" walls. The mixture phase diagrams in the pores are shown in Fig. 4. The influence of the walls on the liquid phase composition is rather weak, while the shift of the vapor composition is more prominent. As expected, the effect is more for the stronger wall potential. In the case of "weak" wall, the shift of the vapor composition is within the statistical errors for 10.2 nm pore width, while the "strong" wall demonstrates effect on phase diagram for both pore widths.

An important result is the prominent shift of the mixture critical point with the 4.08 nm pore with "weak" potential. The critical pressure rises from around 120 atm in the bulk to about 140 atm in slit pore (Fig. 4). Since rocks usually have higher permeability to single-phase supercritical fluid than to two-phase mixture,

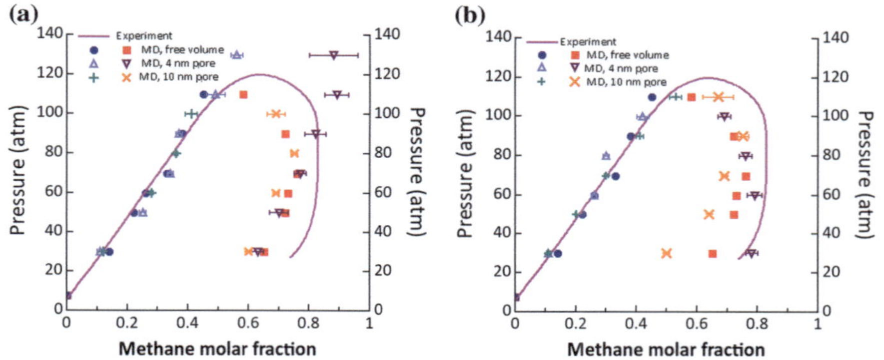

Fig. 4 Phase diagrams for methane + n-butane mixture at 330 K in slit pores with "weak" walls (**a**) and "strong" walls (**b**). *Triangles* 10.2 nm wide pores, crosses: 4.08 nm wide pores. *Circles*, *squares*, and *solid line* are the same as in Fig. 2. The error bars show statistical uncertainties. The error bars lie within the symbols if not shown

the widening of two-phase region may lead to lowering the permeability of nanoporous media.

As of yet, we cannot establish a clear relation between the phase composition shift and the pore width. In the case of the "strong" wall, the vapor composition shift relative to the bulk case has different signs depending on the pore width.

3.3 Phase Diagram of Methane Hydrates

Our results for different water models (TIP4P/Ice, TIP4P/2005, and SPC/E) are shown in Fig. 5.

According to Conde and Vega [31], the TIP4P/Ice model provides the best agreement with the experimental data. Jensen et al. [33] determined the sI melting line by free energy calculations via Monte Carlo method for TIP4P/Ice model, and the agreement of their results is worse than it was found by Conde and Vega (although the LJ potentials for methane were slightly different). Our MD results are in agreement with the data of Jensen et al. This is a strange fact because our results for TIP4P/2005 models are in a fairly good agreement with Conde and Vega. We attribute this discrepancy to the larger interface area of our model (5×5 unit cells compared to 2×2 in Conde and Vega's work). Presumably, smaller interface cross sections can result in larger statistical uncertainty and biased coexisting pressure and temperature values. Our results show that the TIP4P/2005 model gives the better agreement with the experimental coexistence line than the TIP4P/Ice model in the entire pressure range considered. Although TIP4P/2005 coexistence temperatures are systematically 10–20 K lower, the qualitative curve shape reproduces the experimental data quite well.

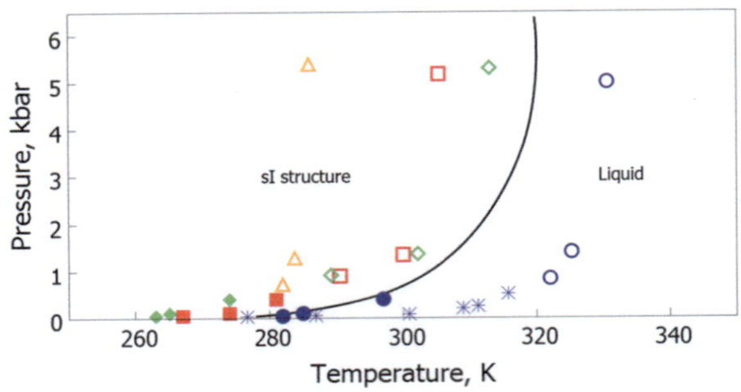

Fig. 5 Methane–water phase diagram. The *solid line* is the experimental [47] three-phase equilibrium curve of methane hydrate. The snowflakes show the result of Jensen et al. [33] for TIP4P/Ice model. The *filled blue, green*, and *red symbols* show the three-phase coexistence points of Conde and Vega [31], and *open symbols* show our results: *Yellow triangles* are for SPC/E, *green diamonds* and *red squares* for TIP4P/2005 with χ = 1:07 and 1.00 in (1), respectively, and *blue circles* for TIP4P/Ice. Our symbols correspond to 5510 unit cell systems. The statistical errors are within the symbols

3.4 Diffusion in Hydrogen Hydrates

We have studied in our previous work [48] the stability areas of the possible structures of the new phase at $\sim$0.5 GPa suggested by experimenters. It turns out that C_0 and sT' structures remain stable in the MD simulations.

We distinguish two characteristic time and length scales of diffusion in both C_0 and sT' structures. On the short timescale, hydrogen molecules move within a single cage (in sT' structure) or channel (in C_0 structure). On longer timescales, molecules jump between cages or channels. The jumps are rather rare events because molecules have to overcome high energy barriers.

Diffusion of guest molecules in C_0 and sT' structures shows prominent anisotropic and anomalous character, i.e., diffusion along different axes occurs at highly different rates, and the mean square displacement (MSD) does not grow linearly on time. Such behavior is probably due to the strong interaction between the framework and guest molecules, since the simulations of gases in metal–organic frameworks with large cage sizes do not reveal anomalous diffusion [49]

Water molecules in the C_0 structure form parallel helical channels oriented along the z-axis. Diffusion of hydrogen in the XY plane occurs therefore due to the jumps of molecules from one channel to another. The time- and ensemble-averaged MSDs of hydrogen molecules along each axis are shown in Fig. 6 in the double logarithmic scale. The asymptotical behavior of MSD curves at long timescales is shown for the lowest and highest of the studied temperatures. The MSD in the z direction (along the axis) is several orders higher than the MSD in the perpendicular plane. The MSDs in the XY plane do not exceed 0.1 nm^2 in 10 ns. This value is less than the square of the distance between channel centers. Therefore, most hydrogen molecules only move along the channel and do not jump between channels on this timescale.

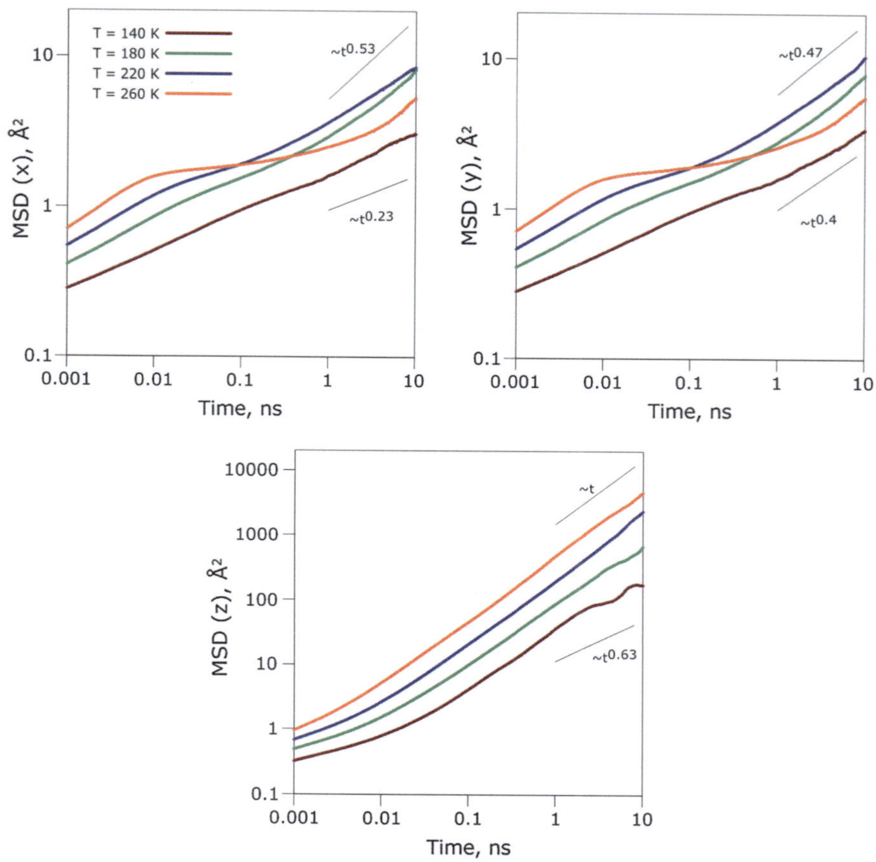

Fig. 6 Time- and ensemble-averaged mean square displacements of hydrogen molecules in the C_0 structure at different temperatures and 100 % cage occupancy

The MSDs along the *x*- and *y*-axes show peculiar behavior at 260 K. The high-temperature MSD curves cross the MSDs at 180 and 220 K. Thus, some inhibition of diffusion takes place at elevated temperatures. Its origin is yet unknown.

The diffusion of hydrogen molecules in the *sT′* structure is even slower (Fig. 7). The MSDs in 10 ns do not exceed 0.04 nm^2 along *x*- and *y*-axes and 0.006 nm^2 along the *z*-axis. The MSDs at 180 and 220 K reach the plateau corresponding to the trapping of molecules within a single polyhedron. Leaving a cage is a very improbable event, so the contribution of such jumps is negligible after ensemble averaging. At 140 K, the diffusion is so slow that only the *z* diffusion curve reaches the plateau. At 260 K, we see the growth of the MSD after a short plateau which corresponds to the jumps of hydrogen molecules between cages. The same behavior is expected for the lower temperatures at longer timescales.

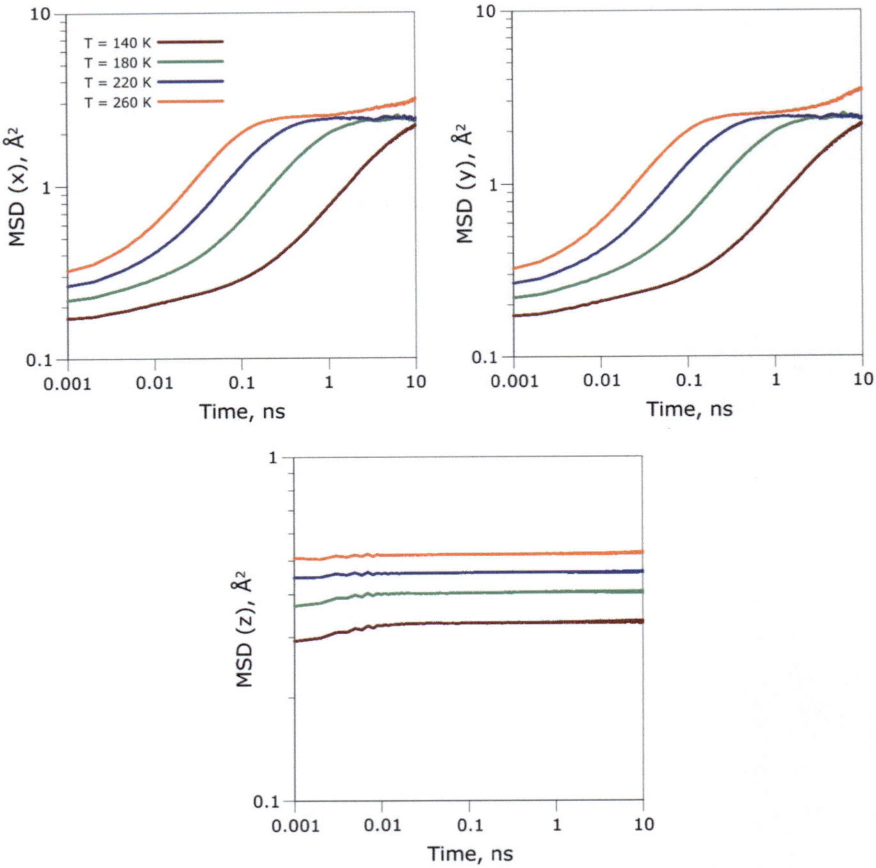

Fig. 7 Time- and ensemble-averaged mean square displacements of hydrogen molecules in the *sT'* structure at different temperatures and 100 % cage occupancy

The analysis of the displacements of the individual molecules shows that the MSDs in the *XY* plane are in fact due to large displacements of only a few molecules (several tens out of several hundred).

One of the important issues is the possibility to reveal the specific mechanisms of subdiffusion. The nonlinear time dependence of mean square displacements appears in different mathematical models, for example, in continuous-time random walk models, fractional Brownian motion, and diffusion on fractals. Sometimes, subdiffusion is a combination of different mechanisms. The more thorough investigation of subdiffusion mechanisms, subdiffusion–diffusion crossover times, diffusion coefficients, and activation energies is the subject of future works.

4 Summary

Two gas extraction problems are formulated to solve with the atomistic modeling. They are related to natural gas condensates and gas hydrates. First steps toward the multiscale modeling are suggested. Examples of molecular dynamics simulations are performed for phase diagrams and diffusion.

- The phase diagram of the test methane + n-butane system is calculated.
- The effect of nanoscale porosity on the test phase diagram is considered. Pore walls are shown to have more effect on the equilibrium vapor composition than on methane solubility. The effect of the critical point shift in nanopores is demonstrated.
- Three-phase coexistence lines are calculated for *sI* methane hydrate using different water models.
- Possible structures of hydrogen clathrate hydrates are refined at high pressures: C_0 and *sT'*.
- Anomalous diffusion of hydrogen molecules is analyzed in these structures, which is determined by the geometry features of the water framework. Diffusion of hydrogen molecules in the new C_0 and *sT'* hydrogen clathrate structures is also analyzed. Mean square displacement analysis shows that diffusion is anisotropic and anomalous at nanosecond timescale.

Acknowledgments The work is supported by the Russian Science Foundation grant 14-50-00124. The authors are thankful to prof. V.M. Zaichenko, who paid our attention to connection of our nucleation study with natural gas condensates modeling and to Drs V.V. Kachalov and V.M. Torchinskii for their interest to the work and valuable discussions.

References

1. Zaichenko, V.M., Maikov, I.L., Torchinskii, V.M., Shpil'rain, E.E.: Simulation of processes of filtration of hydrocarbons in a gas-condensate stratum. High Temp. **47**, 669-674 (2009)
2. Direktor, L.B., Zaichenko, V.M., Maikov, I.L., et al.: Theoretical and experimental studies of hydrodynamics and heat exchange in porous media. High Temp. **48**, 887–895 (2010)
3. Sage, B.H., Hicks, B.L., Lacey, W.N.: Phase equilibria in hydrocarbon systems. The methane-n-butane system in the two-phase region. Ind. Eng. Chem. **32**, 1085 (1940)
4. Muhlbauer, A.: Phase Equilibria: Measurement and Computation. CRC press (1997)
5. Kahre, L.C.: Low-temperature K data for methane-n-butane. J. Chem. Eng. Data **19**, 67–71 (1974)
6. Elliott, D.G., Chen, R.J.J., Chappelear, P.S., Kobayashi, R.: Vapor-liquid equilibrium of methane-n-butane system at low temperatures and high pressures. J. Chem. Eng. Data **19**, 71–77 (1974)
7. Norman, G.E., Stegailov, V.V.: Stochastic theory of the classical molecular dynamics method. Math. Models Comput. Simul. **5**, 305–333 (2013)
8. Rapaport, D.C.: The art of molecular dynamics simulation, 2nd edn. Cambridge University Press (2004)

9. Frenkel, D., Smit, B.: Understanding molecular simulation: from algorithms to applications. Academic Press (2002)
10. Norman, G.E., Filinov, V.S.: Investigation of phase transitions by a Monte-Carlo method. High Temp. **7**, 216–222 (1969)
11. Panagiotopoulos, A.Z.: Direct determination of phase coexistence properties of fluids by Monte Carlo simulations in a new ensemble. Mol. Phys. **61**, 813–826 (1987)
12. Kofke, D.A., Glandt, E.D.: Monte Carlo simulation of multicomponent equilibria in a semigrand canonical ensemble. Mol. Phys. **64**, 1105–1131 (1988)
13. Mehta, M., Kofke, D.A.: Coexistence diagrams of mixtures by molecular simulation. Chem. Eng. Sci. **49**, 2633–2645 (1994)
14. Kaneko, T., Mima, T., Yasuoka, K.: Phase diagram of Lennard-Jones fluid confined in slit pores. Chem. Phys. Lett. **490**, 165–171 (2010)
15. Fomin, YuD: Molecular dynamics simulation of benzene in graphite and amorphous carbon slit pores. J. Comput. Chem. (2013). doi:10.1002/jcc.23429
16. Fomin, YuD, Tsiok, E.D., Ryzhov, V.N.: The behavior of benzene confined in single wall carbon nanotube. J. Comput. Chem. (2015). doi:10.1002/jcc.23872
17. Fomin, YuD, Tsiok, E.D., Ryzhov, V.N.: The behavior of cyclohexane confined in slit carbon nanopore. J. Chem. Phys. **143**, 184702 (2015)
18. Rudyak, V.Ya., Belkin, A.A., Egorov, V.V., Ivanov, D.A.: About fluids structure in microchannels. Int. J. Multiphys. **5**, 145–155 (2011)
19. Rudyak, V.Ya., Belkin, A.A.: Fluid viscosity under confined conditions. Doklady Phys. **59**, 604–606 (2014)
20. Johnston, K., Harmandaris, V.: Properties of benzene confined between two Au(111) surfaces using a combined density functional theory and classical molecular dynamics approach. J. Phys. Chem. C **115**, 14707–14717 (2011)
21. Moustafa, S.G., Schulz, A.J., Kofke, D.A.: Effects of finite size and proton disorder on lattice-dynamics estimates of the free energy of clathrate hydrates. Ind. Eng. Chem. Res. **54**, 4487–4496 (2015)
22. Skripov, V.P., Faizullin, M.Z.: Crystal-Liquid-Gas Phase Transitions and Thermodynamic Similarity. Wiley-VCH, Berlin-Weinheim (2006)
23. Strauss, H.L., Chen, Z., Loong, C.-K.: The diffusion of H_2 in hexagonal ice at low temperatures. J. Chem. Phys. **101**, 7177 (1994)
24. Ildyakov, A.V., Manakov, A.Y.: Solubility of hydrogen in ice Ih at pressures up to 8 MPa. Int. J. Hydrogen Energy **39**, 18958–18961 (2014)
25. Alavi, S., Ripmeester, J.A.: Hydrogen-gas migration through clathrate hydrate cages. Angew. Chem. Int. Ed. Engl. **46**, 6102–6105 (2007)
26. Frankcombe, T.J., Kroes, G.-J.: Molecular dynamics simulations of type-sII hydrogen clathrate hydrate close to equilibrium conditions. J. Phys. Chem. C **111**, 13044 (2007)
27. Iwai, Y., Hirata, M.: Molecular dynamics simulation of diffusion of hydrogen in binary hydrogen–tetrahydrofuran hydrate. Mol. Simul. **38**, 333–340 (2012)
28. Gorman, P.D., English, N.J., MacElroy, J.M.D.: Dynamical cage behaviour and hydrogen migration in hydrogen and hydrogen-tetrahydrofuran clathrate hydrates. J. Chem. Phys. **136**, 044506 (2012)
29. Cao, H., English, N.J., MacElroy, J.M.D.: Diffusive hydrogen inter-cage migration in hydrogen and hydrogen-tetrahydrofuran clathrate hydrates. J. Chem. Phys. **138**, 094507 (2013)
30. Tung, Y.-T., Chen, L.-J., Chen, Y.-P., Lin, S.-T.: The growth of structure I methane hydrate from molecular dynamics simulations. J. Phys. Chem. B. **114**, 10804–10813 (2010)
31. Conde, M.M., Vega, C.: Determining the three-phase coexistence line in methane hydrates using computer simulations. J. Chem. Phys. **133**, 064507 (2010)
32. Abascal, J.L.F., Sanz, E., García Fernández, R., Vega, C.: A potential model for the study of ices and amorphous water: TIP4P/Ice. J. Chem. Phys. **122**, 234511 (2005)
33. Jensen, L., Thomsen, K., von Solms, N., et al.: Calculation of liquid water−hydrate−methane vapor phase equilibria from molecular simulations. J. Phys. Chem. B **114**, 5775–5782 (2010)

34. Smirnov, G.S., Stegailov, V.V.: Melting and superheating of sI methane hydrate: molecular dynamics study. J. Chem. Phys. **136**, 044523 (2012)
35. Abascal, J.L.F., Vega, C.: A general purpose model for the condensed phases of water: TIP4P/2005. J. Chem. Phys. **123**, 234505 (2005)
36. Martin, M.G., Siepmann, J.I.: Transferable potentials for phase equilibria. 1. united-atom description of n-alkanes. J. Phys. Chem. B **102**, 2569–2577 (1998)
37. Cornell, W.D., Cieplak, P., Bayly, C.I., et al.: A second generation force field for the simulation of proteins, nucleic acids, and organic molecules. J. Am. Chem. Soc. **117**, 5179–5197 (1995)
38. Chen, B., Siepmann, J.I.: Transferable potentials for phase equilibria. 3. explicit-hydrogen description of normal alkanes. J. Phys. Chem. B **103**, 5370–5379 (1999)
39. Tuckerman, M., Berne, B.J., Martyna, G.J.: Reversible multiple time scale molecular dynamics. J. Chem. Phys. **97**, 1990–2001 (1992)
40. Hoover, W.G.: Canonical dynamics: equilibrium phase-space distributions. Phys. Rev. A **31**, 1695 (1985)
41. Shinoda, W., Shiga, M., Mikami, M.: Rapid estimation of elastic constants by molecular dynamics simulation under constant stress. Phys. Rev. B **69**, 134103 (2004)
42. Berendsen, H.J.C., Grigera, J.R., Straatsma, T.P.: The missing term in effective pair potentials. J. Phys. Chem. **91**, 6269–6271 (1987)
43. Docherty, H., Galindo, A., Vega, C., Sanz, E.: A potential model for methane in water describing correctly the solubility of the gas and the properties of the methane hydrate. J. Chem. Phys. **125**, 074510 (2006)
44. Plimpton, S.J.: Fast parallel algorithms for short-range molecular dynamics. J. Comp Phys. **117**, 1–19 (1995)
45. Raitza, T., Reinholz, H., Röpke, G., et al.: Laser excited expanding small clusters: single time distribution functions. Contrib. Plasma Phys. **49**, 496–506 (2009)
46. Morozov, I.V., Kazennov, A.M., Bystryi, R.G., et al.: Molecular dynamics simulations of the relaxation processes in the condensed matter on GPUs. Comp. Phys. Comm. **182**, 1974–1978 (2011)
47. Dyadin, Y.A., Aladko, E.Y.: In: Monfort, J. (ed.) Proceedings of the Second International Conference on Natural Gas Hydrates, pp. 67–70 (1996)
48. Smirnov, G.S., Stegailov, V.V.: Toward determination of the new hydrogen hydrate clathrate structures. J. Phys. Chem. Lett. **4**, 3560–3564 (2013)
49. Borah, B., Zhang, H., Snurr, R.Q.: Diffusion of methane and other alkanes in metal-organic frameworks for natural gas storage. Chem. Eng. Sci. **124**, 135–143 (2015)

Atomistic Simulations of CO_2 During "Trapdoor" Adsorption onto Na-Rho Zeolite

Nathan Bamberger and Daniela Kohen

Abstract Behavior of CO_2 within Na-Rho was studied using atomistic simulations. This zeolite is known to experience a phenomenon called "cation gating" which allows carbon dioxide but not other sorbents to permeate the zeolite, giving rise to very high adsorption selectivities for CO_2. Our goal is to provide further insight into the reasons behind this intriguing phenomenon. We show that CO_2's favorable electrostatic interactions with the zeolite framework result in preferential binding in the opening of the channels between cages. This leads us to suggest a novel mechanism to explain carbon dioxide's unique "gate opening behavior" in which this preference for binding inside the "gate" allows CO_2 to "squeeze" by the gate-keeping cation as it moves around slightly due to thermal fluctuations. This proposed mechanism is distinct from a previously proposed mechanism in which carbon dioxide mediates the displacement of gatekeeping cations via electrostatic interactions and may be in better agreement with experimental evidence.

Keywords Zeolites · Cations · Trapdoor · Gating · RHO

1 Introduction

Zeolites and metal-organic frameworks are two fascinating classes of microporous adsorbents with potential applications in separation processes, catalysis, and gas storage [1–4]. In particular, these materials have received a lot of attention due to their potential ability to reduce greenhouse gas emissions through carbon-capture schemes [5, 6].

Both families of materials have advantages and disadvantages for this application, but zeolites are particularly attractive since they are already industrially synthesized, applied in large-scale processes, and can have good stability in the presence of water and other impurities. Framework structure, composition, and

N. Bamberger · D. Kohen (✉)
Chemistry Department, Carleton College, Northfield, MN 55057, USA
e-mail: dkohen@carleton.edu

R.Q. Snurr et al. (eds.), *Foundations of Molecular Modeling and Simulation*,
Molecular Modeling and Simulation, DOI 10.1007/978-981-10-1128-3_10

location of extra-framework cations strongly influence carbon dioxide uptake in zeolites [7]. These factors also affect selectivity, which is just as important for a carbon-capture candidate as being able to strongly adsorb CO_2.

Related zeolite Rho (RHO) materials with a Si/Al framework ratio of 4.5 have shown both good CO_2 uptake and high CO_2 selectivity with respect to small molecules such as CH_4, N_2, and ethane [8–10]. As with other zeolites containing small pores connecting reasonably large cavities [11], RHO materials have window dimensions close to the kinetic diameter of the relevant gases and cage sizes that facilitate interaction with the adsorbing molecules. In addition to these characteristics, many univalent cation-exchanged zeolite Rho materials have extra-framework cations that block the entrances to the narrow pores connecting cages [9, 10]. Such materials experience a phenomenon called "cation gating" which allows carbon dioxide but not other sorbents to permeate the zeolite, giving rise to very high adsorption selectivities for CO_2. This complex behavior is a consequence of the siting and movement of these extra-framework cations but also of the strong cation-dependent structural flexibility of the Rho structure. These materials expand to accommodate carbon dioxide (and presumably no other adsorbate), but the extent of the change depends on the nature of the cation.

In this work, we focus on fully exchanged Na-Rho, the most studied Rho structure and the most promising for practical applications due to its adsorption properties and costs. Na-Rho has been extensively studied by Lozinska et al. [9, 10], who found that although it is a flexible zeolite, it retains its symmetry when loaded with 1 bar of CO_2 and distorts and expands less than other Rho materials. Lozinska et al. also performed careful in situ XRD structural studies and IR spectroscopy of carbon dioxide adsorption within this material. These studies as well as others in related Rho materials led these authors to propose a mechanism for cation gating in which cations in window sites interact strongly enough with nearby carbon dioxide molecules that the cations are temporarily displaced to empty sites within nearby α-cages, opening a "trapdoor" that allows adsorbates to diffuse through the zeolite. This intriguing mechanism is likely to be at play in other relevant materials as well. In particular, Webley and coworkers believe that a "trapdoor" mechanism is also responsible for the very high selectivity of carbon dioxide over methane that they have found in chabazite zeolites [12–14].

Because cation gating is fundamentally a molecular scale phenomenon, atomistic simulations are well-suited to help answer some of the questions left open by the experiments of Lozinska et al. due to the time-averaged nature of their data.

In this paper, we therefore present classical molecular simulations that provide detailed microscopic information regarding carbon dioxide's "gated" adsorption within fully exchanged Na-Rho. However, the scope of this work is modest, as we only focus on the behavior of mobile cations and carbon dioxide within a rigid zeolite. This is because, to the best of our knowledge, there are no reliable atomistic potentials that can model a flexible aluminum-substituted zeolite framework. Our goal is to provide further insight into the reasons behind carbon dioxide's ability to adsorb within cation gated zeolites rather than fully describe the system's behavior. In particular, our results suggest that the carbon dioxide, rather than mediating the displacement of gating cations to open the trapdoor, is instead more adept than other sorbents at "squeezing" through the trapdoor as it opens by itself due to thermal fluctuations.

The remainder of the paper is structured as follows. In Sect. 2, we describe our computational methods. Section 3 presents our results and discussion: Sect. 3.1 presents cation radial distribution functions in the presence and absence of carbon dioxide, and Sect. 3.2 describes carbon dioxide and Na^+ preferred sites of adsorption. These two sections provide the rationale for the alternative scenario described in the previous paragraph and set the stage for Sect. 3.3, where we show a suggestive MD simulation of a carbon dioxide entering a "blocked" channel. We conclude in Sect. 4.

2 Methods and Models

All of our atomistic simulations were performed using standard Grand Canonical Monte Carlo (GCMC) and Equilibrium Molecular Dynamics (EMD) simulation methods. The RASPA [15] code was employed. Electrostatic energies were calculated using Ewald summation [16, 17] with a relative error of 10^{-6}. A 12 Å van der Waals cutoff was used for the short-range interactions. Periodic boundary conditions were employed.

In our GCMC simulations, four types of trial moves were used: attempts to translate an adsorbed carbon dioxide or a sodium cation, attempts to insert a new carbon dioxide into the zeolite, attempts to delete an existing carbon dioxide from the zeolite, and attempts to rotate an adsorbed carbon dioxide. Typically, simulations were run for 5×10^6 Monte Carlo cycles (each cycle consisted of max[N, 20] steps where N is the number of moving particles). The first half of these cycles was used for equilibration and was not included in the sampling of the desired thermodynamic properties. MD simulations were performed in the NVT ensemble at 298 K using a Nose–Hoover thermostat to regulate the temperature. The time step in all simulations was 0.5 fs. Each MD simulation started with a Monte Carlo (MC) pre-equilibration (at least 10^6 Monte Carlo moves) followed by MD equilibration (at least 10^6 MD steps). After equilibration, production runs of 10^6 MD steps were performed and used to sample the desired thermodynamic properties. In both, the MD and GCMC simulations, the number of steps (or cycles) was large enough that the results were independent of the number of steps.

In the work presented here, interactions between adsorbed molecules, the negatively charged zeolite framework, and extra-framework cations are modeled using a DFT-derived force field for carbon dioxide in Na-exchanged zeolites. Recently developed by Sholl's group [18] and referred to as CCFF, this potential was obtained using experimental data for zeolite LTA-4A and validated with two other common adsorbents, NaX and NaY. This makes it ideally suited for our purposes as it was designed with the goals of both being accurate and transferable to materials with the same chemical composition as Na-Rho. The CCFF potential

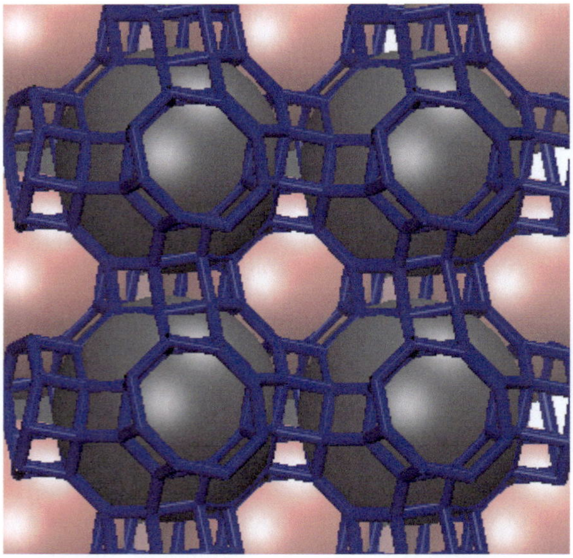

Fig. 1 The structure of zeolite Na-RHO when $P_{CO_2} = 1$ bar [20]. The framework is shown in *blue*. Note the 3-dimensional channel system composed of cavities (α-cages). Each α-cage is connected to six others by D8Rs. This gives rise to two interpenetrated but not interconnected pore systems (shown in *gray* and *pink*)

models CO_2–CO_2 interactions using the well-established EPM2 potential [19]. Within this potential, carbon dioxide is represented by a linear triatomic with fixed bond lengths and bond angles and each atom is described by a charged Lennard-Jones (LJ) center. All other interactions in the system are modeled using DFT-derived parameters. The interaction between carbon dioxide and the rest of the system has two contributions: a Coulombic and a LJ interaction between each pair of atoms. The interaction of each extra-framework cation and framework atom has a Coulombic contribution as well. In addition, the dispersion interaction between each extra-framework cation and the oxygen atoms within the framework is modeled using a Buckingham potential.

As was mentioned in the introduction, the work presented here focuses on behavior within the Na-Rho zeolite (Fig. 1). The structure of zeolite RHO is well known [20]; it has a 3-dimensional channel system composed of one size of cavities (α-cages). Each α-cage is connected to six other α-cages by double 8-ring pores (D8R). This gives rise to two interpenetrated but not interconnected pore systems. When dehydrated, zeolite Na-Rho has $I\bar{4}3m$ symmetry [9]. This zeolite is flexible: when loaded with 1 bar of carbon dioxide, the 8-rings are distorted from circular to elliptical, the α-cages become tetrahedral rather than cubic, and the zeolite expands approximately 2 % (but maintains $I\bar{4}3m$ symmetry) [10].

To the best of our knowledge, there is no available potential that would allow us to describe both the flexibility and the carbon dioxide absorption of this zeolite, and so instead, we model the system with all framework atoms fixed at their crystallographic positions while cations and carbon dioxide molecules are allowed to move. In order to make 1-to-1 comparisons regarding sites and locations within the zeolite, we wanted to use the same rigid zeolite structure for both the simulations with CO_2 and without CO_2, even though as described above the structure is known to change upon addition of CO_2. We chose to use the Na-Rho crystallographic positions corresponding to the zeolite structure when loaded with 1 bar of CO_2 [10] because our primary interest is in interactions involving CO_2. Note that most classical simulations of cation-exchanged zeolites [21, 22] also assume a rigid framework and that the CCFF potential was derived under these conditions as well. Furthermore, studies of methane in flexible LTA zeolite [23] have shown that flexibility is much less important when studying adsorption than when studying diffusion. All qualitative conclusions described in what follows were obtained using both GCMC (simulating adsorption) and MD (simulating diffusion) calculations (unless noted), lending credibility to our approach. However, the absence of a potential that can be used to more accurately study the behavior of carbon dioxide within flexible Na-Rho limits the scope of this work to the qualitative insights provided in the results and conclusion sections.

Before continuing note that when using crystallographic data obtained for Na-Rho loaded with 1 bar of CO_2, our approach is able to accurately reproduce a 298 K experimental isotherm (see Fig. 2). Note that in all the simulations presented in this work, the Si/Al ratio is approximately 4 (9.8 Al and 38.2 Si per unit cell) in order to simulate the material of interest. The positions of the Al atoms are chosen randomly subject to the constraint of Lowenstein's rule [24].

Lozinska et al. have determined that when the ratio of Si/Al $\approx$ 4, Na^+ cations preferentially occupy S8R (single 8 ring) sites and S6R (single 6 ring) sites (see Fig. 3). Our simulations of this system show cation sites that are very similar to the experimental ones in both location and fractional occupancy (see Fig. 3 and Table 1).

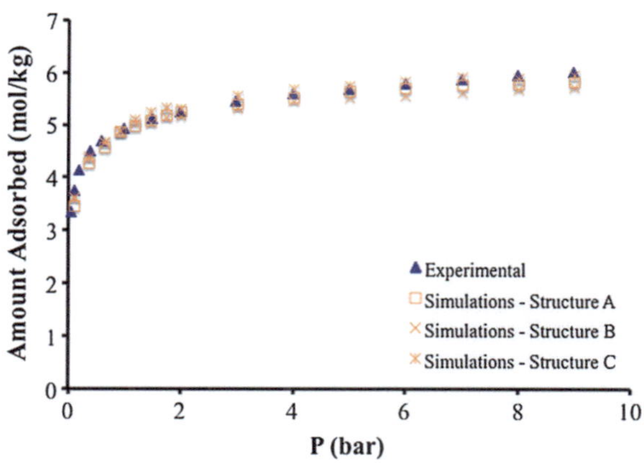

Fig. 2 CO_2 adsorption isotherm within Na-RHO at 298 K. The simulations were performed using a rigid zeolite framework with all atoms at their crystallographic positions. The three simulations shown differ in the random locations of the Al-substitutions within the zeolite. The experimental data are that of Ref. [10]

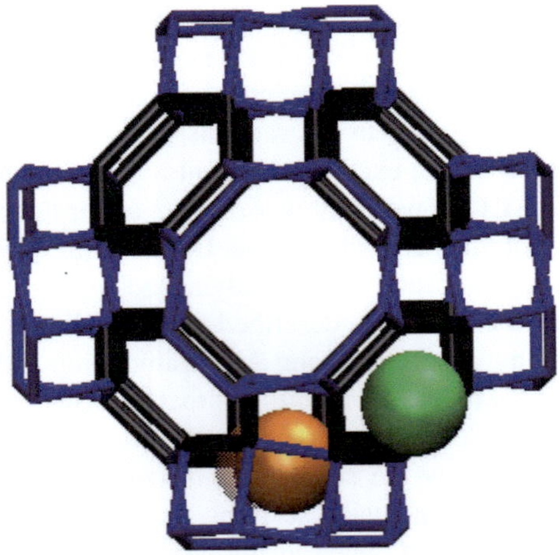

Fig. 3 Cation Sites. *Spheres* showing preferred cation sites: S8R in *orange* and S6R in *green*. The orientation of the cage was chosen to clearly show how the experimental cation site in the S8R (*solid orange sphere*) is not exactly the same as the one found in our simulations (*translucent orange sphere*); although it is not obvious in the figure, the experimental site refines in an off-center position. Note that the experimental S6R site is indistinguishable from the one found in this work. The figure also shows the D8R traced in *blue* and the S6R traced in *black*. Note that in each unit cell, there are 6 D8R sites (and thus 12 S8R) and 8 S6R. The experimental data are that of Ref. [10]

Table 1 Fractional occupancies of cation sites

Sample (P_{CO_2})		Occ. fraction site S8R	Occ. fraction site S6R
Na-RHO (0 bar)	Experimental	0.51	0.43
	Simulations	0.48	0.41
Na-RHO (1 bar)	Experimental	0.5	0.49
	Simulations	0.46	0.40

Experimental data are from Ref. [10]. Note that in our calculations, as in experiments, about half the S8R are blocked by cations whether carbon dioxide is present or not. Lozinska et al. argue that it is likely that cations prefer to occupy S8Rs belonging to different D8Rs. However, our calculations show that is not the case. This could be another shortcoming of the interaction potentials we are using, but without further evidence, this cannot be ascertained

3 Results and Discussion

3.1 Cation Radial Distribution Function

The equilibrium positions of cations in the presence and absence of carbon dioxide were first investigated by calculating radial density probability functions. Figure 4 shows the average of 5 such plots from different GCMC runs where the random Al

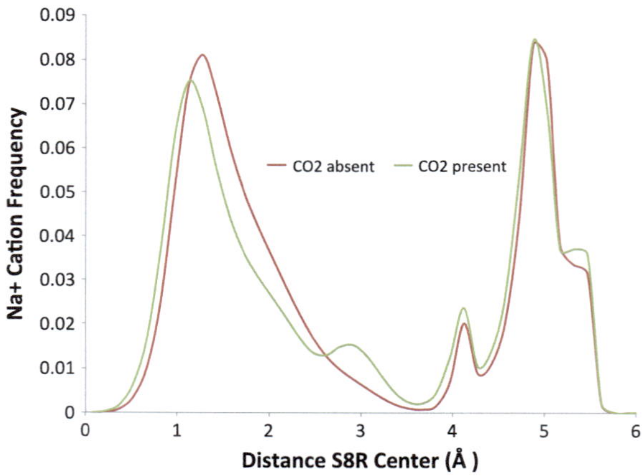

Fig. 4 Cation radial probability density functions. Each line is the average of 5 plots from different GCMC runs where the random Al locations were all the same but the seed numbers were different. Note that the distributions do not differ substantially between when carbon dioxide is present and when it is not. In both cases, there is a peak near zero corresponding to the S8R site and another much further away corresponding to cations on the S6R site. However, while the distribution of cations in the S6R is essentially unchanged, there is a small shift in the distribution of cations in the S8R when carbon dioxide is present: The probability of a cation being between 1–2 Å from the center of the ring is smaller while around 3 Å the probability is larger than in the absence of the adsorbate

locations were all the same, but the seed numbers were different. Note that the qualitative conclusions reached below can be obtained by examining each of these five runs independently or equivalent MD runs, but averaging allows the reader to focus on the important features. Also note that the conclusions are the same if the random Al-substitutions are different. The zero was chosen as the crystallographic center of an S8R. This position was chosen because a cation sitting near the center of the S8R blocks carbon dioxide molecules from also fitting in the plane of the ring. Figure 4 shows that the distributions do not vary much between when carbon dioxide is present and when it is not. In both cases, there is a peak near zero (corresponding to the cations in the S8R site) and another peak 4–6 Å away corresponding to cations in the S6R site. However, there is one difference that is important in the context of this article: While the distribution of cations in the S6R does not change appreciably, there is a shift in the distribution of cations in the S8R when carbon dioxide is present. More specifically, when CO_2 is added to the zeolite the probability of a cation being located 1–2 Å from the center of the ring decreases while at around 3 Å the probability increases. In other words, there is a net movement of some cations away from the center of the 8 rings when CO_2 is introduced into the zeolite, a finding that is in agreement with the experimental observation that CO_2 is not blocked by cations. However, this finding does not address the issue of what causes some cations to move from their blocking positions and allow for the "opening of the gate." We therefore move on to this question in the following sections.

3.2 Preferred Sites of Adsorption

In this section, preferred sites of adsorption for both carbon dioxide and cations will be investigated. These will suggest that the reason for carbon dioxide's "gate opening behavior" has more to do with carbon dioxide's preferred sites within the zeolite than its ability to guide cations out of the way.

Figures 5 and 6 show probability maps for sodium cations and carbon dioxide molecules, respectively, at 298 K. These maps are normalized 3D histograms of particle locations collected every 10 cycles during a GCMC simulation with 5×10^6 production cycles. Qualitatively equivalent results are obtained if data are collected in an EMD simulation. The maps in Fig. 5 correspond to cation locations when no carbon dioxide is present, but almost identical ones are obtained in the presence of 1 bar of CO_2. This demonstrates that only limited cation rearrangement takes place when carbon dioxide is adsorbed, just as experiments have suggested [9]. In Fig. 5a, the scale is such that even locations that are very infrequently visited are shown. This map reveals that cations explore within their site and thus must somehow be mobile. The extent to which this mobility is poor is underscored in Fig. 5b, where only the region within the 0.1*max probability contour is shown.

Figure 6 shows that, as expected, the carbon dioxide molecules explore a large region of the zeolite; the figure highlights the cages and the narrow passages between then. This plot suggests that the highest probability of finding an adsorbed CO_2 is at the entrances of the narrow channels connecting cages (i.e., in the S8R). This finding is in line with previous work within silica-only zeolites ITQ-3 and ZK4 (an LTA equivalent) [25, 26] where we have shown that carbon dioxide, but not nitrogen, strongly adsorbs in narrow pores between cages. Carbon dioxide adsorption in the narrow pores takes on increased significance in the context of this paper because to enter these pores the CO_2 must pass the cation that is "gating" the S8R.

Figure 7a shows that carbon dioxide has two preferred adsorption sites: one in the S8R, as mentioned previously, and the other, near the walls of the α-cage. These locations are in agreement with findings by Lozinska and coworkers [10] who were also able to locate carbon dioxide molecules in two sites, one within the window region and another within the cage. Figure 7 was obtained by using appropriate symmetry operations to collapse probability anywhere in the simulation cell around a location at the geometric center of the D8R. This effectively moves all the probability around any of the D8Rs in the simulation cell to the vicinity of the one shown, highlighting the role of this region. Figure 7b is an equivalent plot for the sodium cations. Figure 7 shows how both cations and carbon dioxide have a preferred site of adsorption near the center of the S8R.

In their work, Lozinska et al. suggest that blocking cations undergo a quick CO_2-mediated migration from a window site to an empty S6R site, "opening the gate" for a brief period of time and allowing CO_2 to diffuse through. They suggest that weaker electrostatic interactions between cations and other adsorbents prevent this mechanism from taking place with molecules such as CH_4. The plots in Fig. 7 suggest an alternative explanation for carbon dioxide's ability to permeate the

(a)

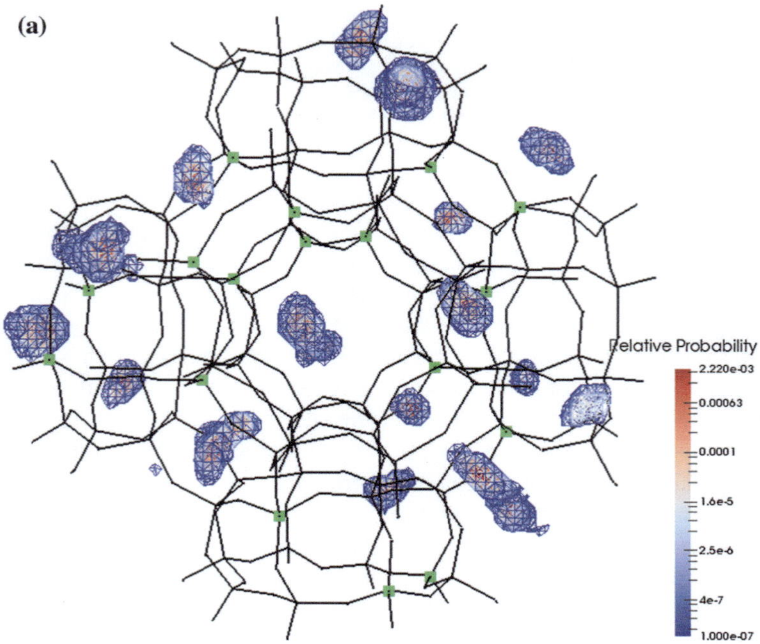
Relative Probability
2.220e-03
0.00063
0.0001
1.6e-5
2.5e-6
4e-7
1.000e-07

(b)

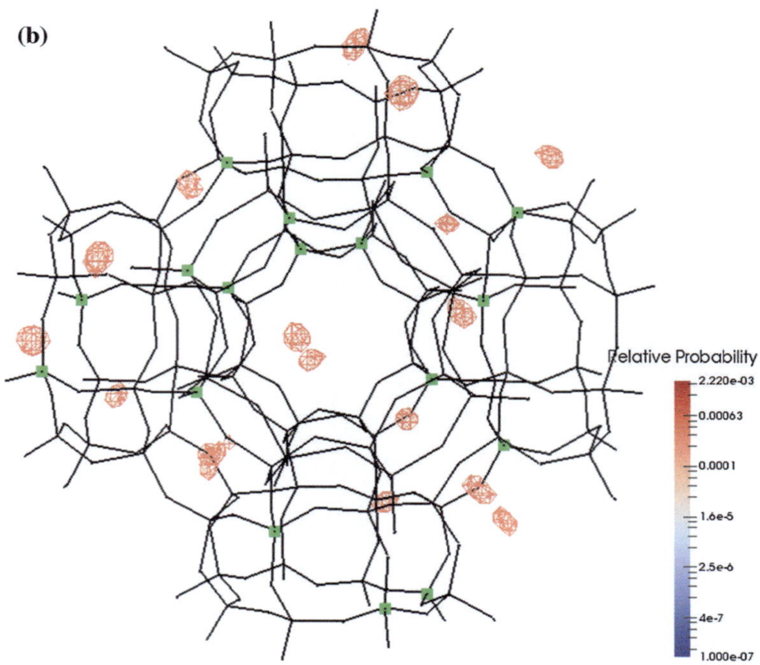
Relative Probability
2.220e-03
0.00063
0.0001
1.6e-5
2.5e-6
4e-7
1.000e-07

◀ **Fig. 5** Probability map for sodium cations. The figure is centered in the middle of a cage, with six D8Rs surrounding it. The framework is shown in *black* with the Al atoms in *green*. **a** The probability scale is such that even locations that are very infrequently visited are shown in *blue*. These *blue* regions reveal that the cations explore within their site and are thus to some extent mobile. **b** Most visited locations (the 0.1*max probability contour is shown)

zeolite: perhaps the gating cation periodically moves around the S8R by itself due to thermal motion, and CO_2 is simply more inclined than other molecules to "squeeze" by into the S8R when this happens due to its natural affinity for this site. In other words, it is not so much that the carbon dioxide molecule opens the gate by interacting with the "gatekeeper" sodium, but rather that carbon dioxide is able to take advantage of a "wandering gatekeeper." In the context of this alternative explanation, Fig. 4 can be understood as showing the cation distribution changing due to competition with the CO_2 for the ring site.

Within this alternative explanation, a methane molecule, for example, would not be able to take advantage of the cation's thermal motion because in the absence of strong electrostatic interactions with the pore it might lack carbon dioxide's preference for an S8R site. At this moment, there is no methane potential that would allow us to further confirm this hypothesis, but to explore the plausibility of this explanation we ran simulations in which we artificially set CO_2's partial charges to

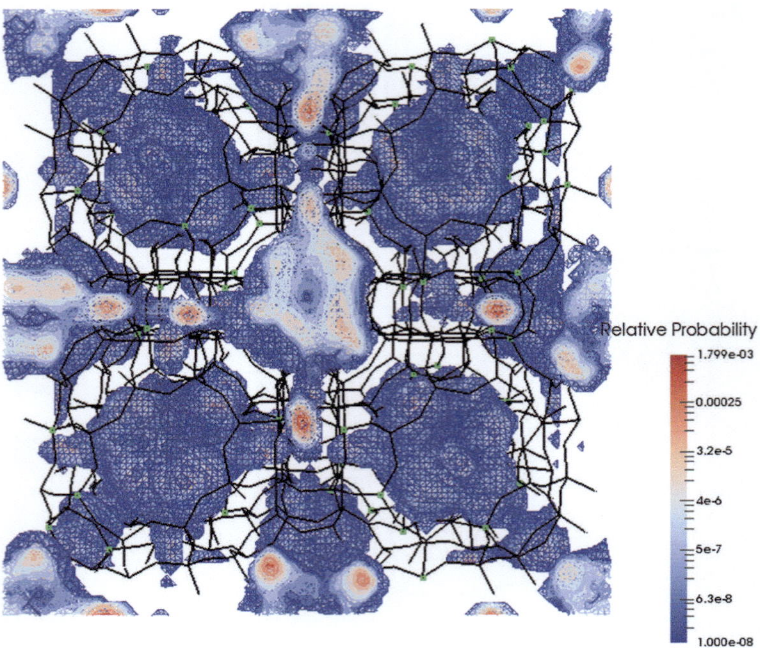

Fig. 6 Probability map for carbon dioxide. The figure shows a 2 × 2 × 2 simulation cell. The framework is shown in *black* with the Al atoms in *green*. The figure highlights the cages and the narrow passages in between then

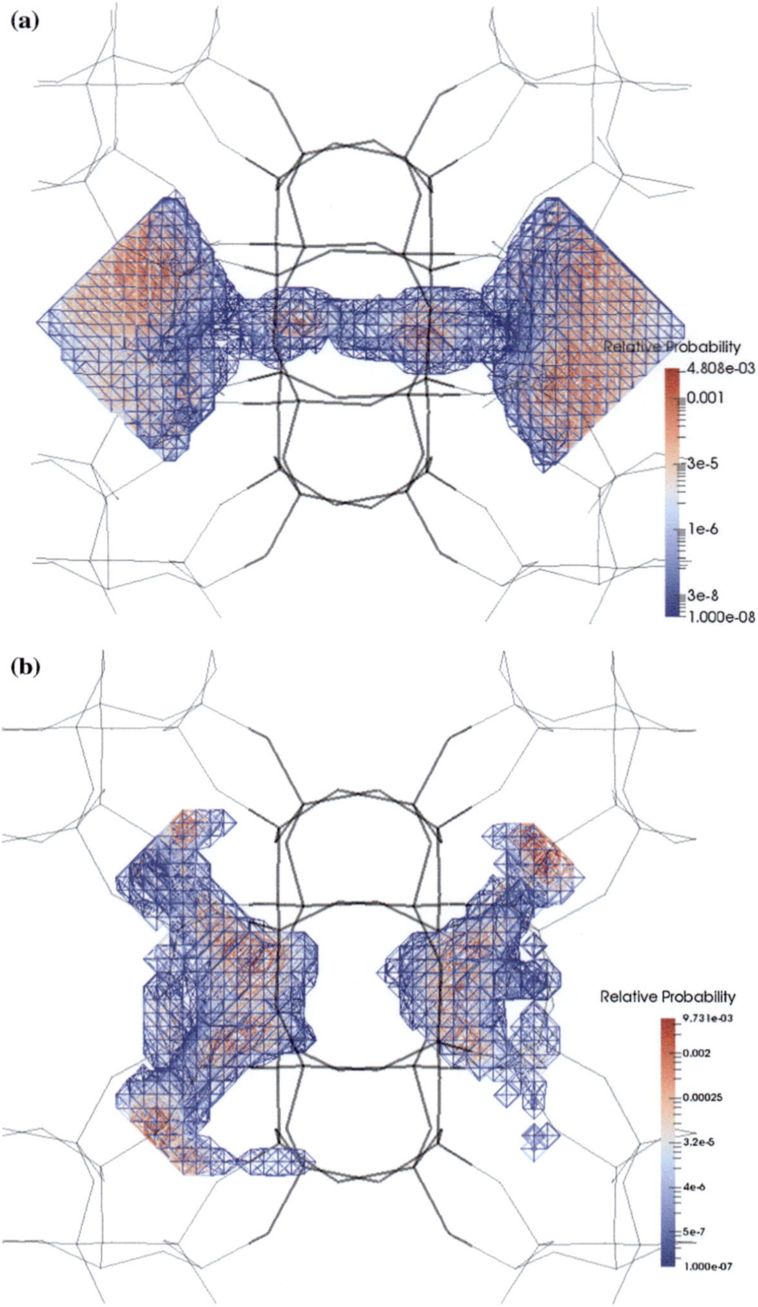
(a)
Relative Probability
4.808e-03
0.001
3e-5
1e-6
3e-8
1.000e-08
(b)
Relative Probability
9.731e-03
0.002
0.00025
3.2e-5
4e-6
5e-7
1.000e-07

◀ **Fig. 7** Probability maps around a D8R. *Thicker lines* highlight the D8R. These plots were obtained by using appropriate symmetry operations to collapse probability anywhere in the simulation cell around a location at the geometric center of the D8R. This effectively collapses all the probability around any of the D8Rs in the simulation cell to the vicinity of the one shown, highlighting the role of this region. **a** Map for carbon dioxide. Note the areas of higher probability in the middle of each S8R and near the walls of the α-cage. The map does not include CO_2 probability corresponding to empty D8R sites. **b** Map for sodium cations. Note the areas of higher probability at the sites mentioned, in the S8R and in the S6R

zero. In previous work within silica-only zeolite [25, 27], we have used this strategy to show that when carbon dioxide is modeled using only dispersive forces, it no longer strongly adsorbs in the narrow pores between cages and this region becomes a barrier to diffusion rather than an adsorption site. Under these conditions, carbon dioxide's preferred sites become quite similar to those of nitrogen (the other adsorbent studied in those articles). Figure 8 is the equivalent of Fig. 7a for when carbon dioxide's partial charges are set to zero. Note how in this case the middle of each S8R is no longer a preferred location. This suggests that a molecule without a significant quadrupole would not "squeeze" into the S8R, and thus, its diffusion would be blocked by the cation in the S8R. This might explain why other small molecules are not able to diffuse within this material [8–10] while carbon dioxide is.

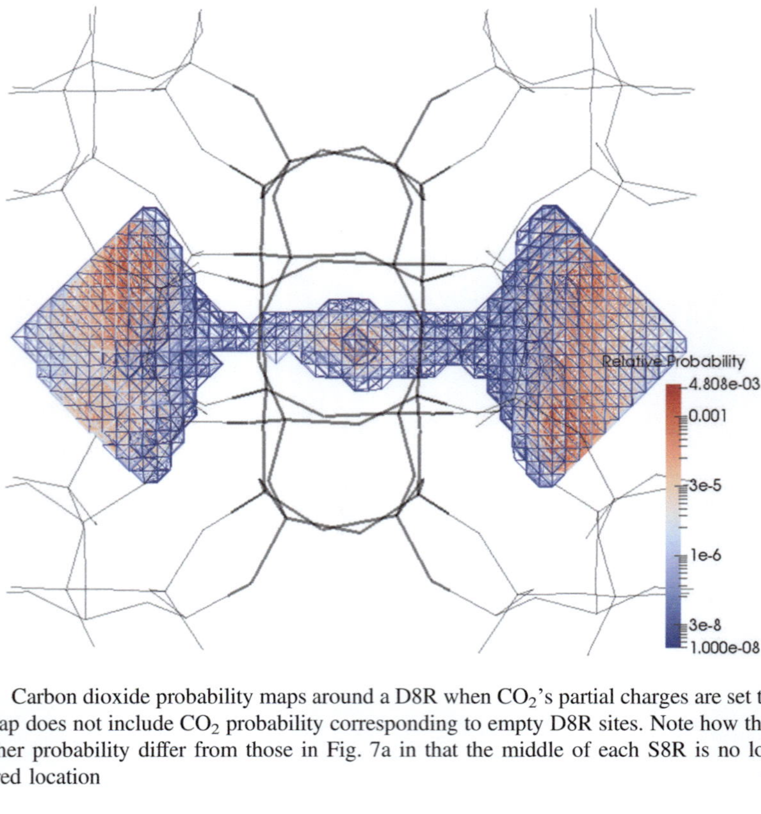

Fig. 8 Carbon dioxide probability maps around a D8R when CO_2's partial charges are set to zero. The map does not include CO_2 probability corresponding to empty D8R sites. Note how the areas of higher probability differ from those in Fig. 7a in that the middle of each S8R is no longer a preferred location

3.3 An MD Trajectory Showing a CO_2 Entering a "Gated" Ring

A carbon dioxide entering a cation-blocked narrow channel between two α-cages (the D8R) is likely a rare event. Given the importance of such an event in the context of this work, many MD simulations were searched in order to find one. Figure 9 shows a carbon dioxide molecule entering a D8R that is blocked by a cation. Figure 9a is a snapshot, Fig. 9b traces the motion of the carbon dioxide, and Fig. 9c traces the cation (a 0.5-ns movie showing this event is available as supplemental information). This figure shows how little the cation moves as the carbon dioxide

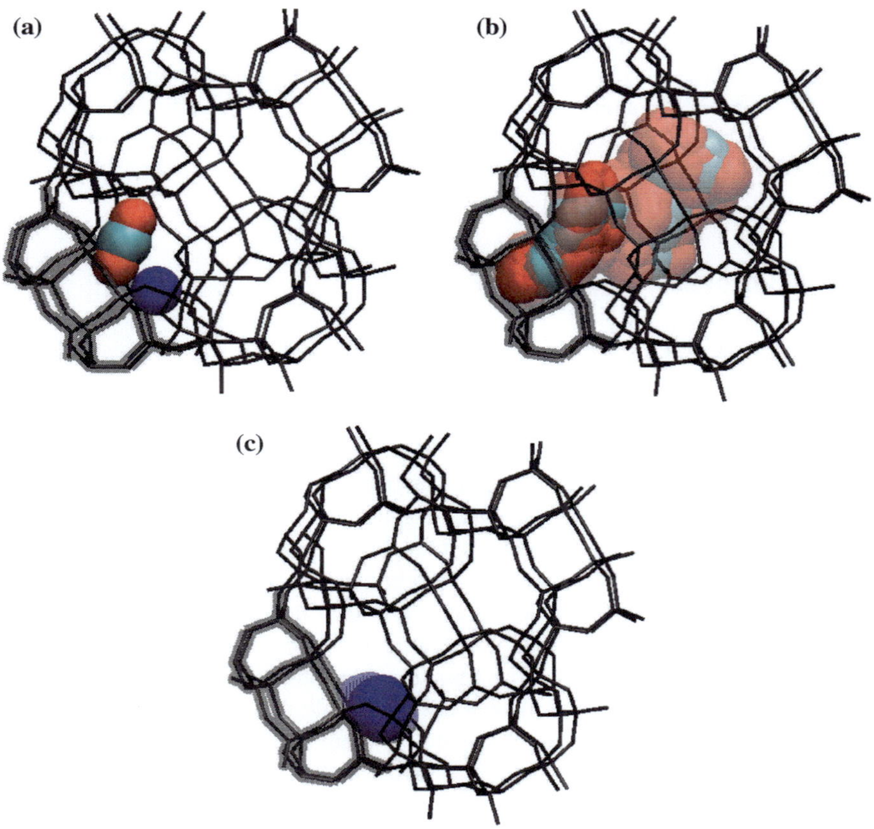

Fig. 9 A carbon dioxide molecule entering a D8R blocked by a cation. Only the relevant adsorbate molecule and cation are shown in the figure. The blocked D8R is highlighted. **a** A snapshot. **b** Multiple frames showing the motion of the CO_2. *Solid colors* show 0.5 ps before and after the frame shown in (**a**) while translucent shows 50 ps before and after. **c** Multiple frames showing the very narrow range of motion of the blocking cation. Frames 0.5 ps before and after the frame in (**a**) are shown in *solid blue*, frames 50 ps before and after are shown in *blue stripes*, and frames 200 ps before and after are shown as translucent *blue*

enters the ring, and in particular that while our simulations show that when the cation and carbon dioxide are close to each other electrostatic interactions cause them to interact strongly, we have found no evidence that the cation needs to leave the S8R site in order to allow for a carbon dioxide molecule to enter the "gate." This finding is significant in the context of CO_2/CH_4 breakthrough experiments performed by Palomino et al. [8] on zeolite Na-RHO. These experiments showed that while CO_2 is retained, methane passes with practically no retention. If the cation truly does leave the narrow channel via a CO_2-facilitated jump from a S8R site to a S6R site (as in the mechanism suggested by Lozinska et al.), even for a short time, then molecules other than carbon dioxide might be able to pass through the gate. On the other hand, a cation that only moves slightly away from the S8R could completely block methane from entering the zeolite. Our proposed mechanism may therefore be in better agreement with the Palomino results than that of Lozinska et al.

It is important to point out, though, that our mechanism and Lozinska's are not mutually exclusive: It is entirely possible that CO_2 being better at opening the gate and CO_2 being better at entering the gate when it opens by itself (due to thermal fluctuations) both contribute to carbon dioxide's ability to diffuse through the zeolite while other molecules cannot.

4 Conclusions

We have used molecular simulations to examine the behavior of CO_2 within Na-Rho zeolite focusing on the manner by which pore blocking sodium cations allow this adsorbate to diffuse within the material. Experiments have shown that while carbon dioxide can explore this zeolite (and other related zeolites), other small gas molecules such as nitrogen and methane cannot, effectively making Na-RHO zeolite a very attractive candidate for practical separations. While this highly selective trapdoor adsorption is thought to be a consequence of both the cation behavior and the framework flexibility, we focus in this work only on the former. Despite this shortcoming, our work identifies a novel understanding of the mechanism at play: rather than coaxing the cation off the blocking position by interacting via electrostatic forces, a carbon dioxide competes with the cation for the position at the entrance of the channel and so is able to squeeze by as the cation moves around its adsorption site (the whole system is at room T). We show that carbon dioxide has a preferred site at the S8R, which disappears when electrostatic forces are artificially ignored. This suggest that gases that do not possess a quadrupole cannot diffuse through S8Rs because they have no energetic reason for entering the narrow channel when then the cation is not quite blocking the S8R.

Our work highlights the need for reliable atomistic potentials for other cations and other adsorbates that would allow this mechanism to be studied further. Potentials that would allow for a flexible zeolite are also needed. We believe the potential used here is adequate to shed light onto the behavior, but its inability to describe motion of the framework is a significant shortcoming. Experimentally it has been shown that carbon dioxide but not other gases can penetrate and thereby distort the zeolite. Our work suggests carbon dioxide's preference for the S8R might influence (and even drive) this geometry change, but the available potentials do not allow us to investigate this hypothesis further. Further improvements in the available force fields would allow for better understanding of materials with doorkeeping cations and their interactions with adsorbates with and without a quadrupole and thus aid the search for microporous materials uniquely suited to practical CO_2 separations.

Acknowledgments N.B. and D.K. gratefully acknowledge the Petroleum Research Fund (PRF# 51765-UR5) and National Science Foundation (CHE-1039925) for computing resources and stipend support to carryout this research.

References

1. Choi, S., Drese, J.H., Jones, C.W.: Adsorbent materials for carbon dioxide capture from large anthropogenic point sources. Chemsuschem **2**, 796–854 (2009)
2. Keskin, S., Sholl, D.S.: Efficient methods for screening of metal organic framework membranes for gas separations using atomically detailed models. Langmuir **25**, 11786–11795 (2009)
3. Yazaydin, A.O., Snurr, R.Q., Park, T.H., Koh, K., Liu, J., LeVan, M.D., Benin, A.I., Jakubczak, P., Lanuza, M., Galloway, D.B., Low, J.J., Willis, R.R.: Screening of metal-organic frameworks for carbon dioxide capture from flue gas using a combined experimental and modeling approach. J. Am. Chem. Soc. **131**, 18198–18199 (2009)
4. Di Biase, E., Sarkisov, L.: Systematic development of predictive molecular models of high surface area activated carbons for adsorption applications. Carbon **64**, 262–280 (2013)
5. D'alessandro, D.M., Smit, B., Long, J.R.: Carbon dioxide capture: prospects for new materials. Angew. Chem. Int. Edit. **49**, 6058–82 (2010)
6. Pera-Titus, M.: Porous inorganic membranes for CO_2 capture: present and prospects. Chem. Rev. **114**, 1413–1492 (2014)
7. Grajciar, L., Cejka, J., Zukal, A., Arean, C.O., Palomino, G.T., Nachtigall, P.: Controlling the adsorption enthalpy of CO_2 in zeolites by framework topology and composition. Chemsuschem **5**, 2011–2022 (2012)
8. Palomino, M., Corma, A., Jorda, J.L., Rey, F., Valencia, S.: Zeolite Rho: a highly selective adsorbent for CO_2/CH_4 separation induced by a structural phase modification. Chem. Commun. **48**, 215–217 (2012)
9. Lozinska, M.M., Mangano, E., Mowat, J.P.S., Shepherd, A.M., Howe, R.F., Thompson, S.P., Parker, J.E., Brandani, S., Wright, P.A.: Understanding carbon dioxide adsorption on univalent cation forms of the flexible zeolite rho at conditions relevant to carbon capture from flue gases. J. Am. Chem. Soc. **134**, 17628–17642 (2012)
10. Lozinska, M.M., Mowat, J.P.S., Wright, P.A., Thompson, S.P., Jorda, J.L., Palomino, M., Valencia, S., Rey, F.: Cation gating and relocation during the highly selective "trapdoor" adsorption of CO_2 on univalent cation forms of zeolite rho. Chem. Mater. **26**, 2052–2061 (2014)
11. Cheung, O., Hedin, N.: Zeolites and related sorbents with narrow pores for CO_2 separation from flue gas. RSC Adv. **4**, 14480–14494 (2014)
12. De Baerdemaeker, T., De Vos, D.: Gas separation trapdoors in zeolites. Nat. Chem. **5**, 89–90 (2013)

13. Shang, J., Li, G., Singh, R., Gu, Q.F., Nairn, K.M., Bastow, T.J., Medhekar, N., Doherty, C. M., Hill, A.J., Liu, J.Z., Webley, P.A.: Discriminative separation of gases by a "molecular trapdoor" mechanism in chabazite zeolites. J. Am. Chem. Soc. **134**, 19246–19253 (2012)
14. Shang, J., Li, G., Singh, R., Xiao, P., Liu, J.Z., Webley, P.A.: Determination of composition range for "molecular trapdoor" effect in chabazite zeolite. J. Phys. Chem. C **117**, 12841–12847 (2013)
15. Dubbeldam, D., Calero, S., Ellis, D.E., Snurr, R.Q.: RASPA: molecular simulation software for adsorption and diffusion in flexible nanoporous materials. Mol. Simul. (2015)
16. Allen, M.P., Tildesley, D.J.: Computer Simulations of Liquids. Oxford Science Publications (1994)
17. Frenkel, D., Smit, B.: Understanding Molecular Simulation: From Algorithms to Applications. Academic Press, London (1996)
18. Fang, H.J., Kamakoti, P., Ravikovitch, P.I., Aronson, M., Paur, C., Sholl, D.S.: First principles derived, transferable force fields for CO_2 adsorption in Na-exchanged cationic zeolites. Phys. Chem. Chem. Phys. **15**, 12882–12894 (2013)
19. Harris, J.G., Yung, K.H.: Carbon dioxides liquid-vapor coexistence curve and critical properties as predicted by a simple molecular-model. J. Phys. Chem.-Us. **99**, 12021–4 (1995)
20. Robson, H.E., Shoemake, D.P., Ogilvie, R.A., Manor, P.C.: Synthesis and crystal-structure of zeolite rho—new zeolite related to linde type-A. Adv. Chem. Ser. 106–15 (1973)
21. Calero, S., Dubbeldam, D., Krishna, R., Smit, B., Vlugt, T.J.H., Denayer, J.F.M., Martens, J. A., Maesen, T.L.M.: Understanding the role of sodium during adsorption: a force field for alkanes in sodium-exchanged faujasites. J. Am. Chem. Soc. **126**, 11377–11386 (2004)
22. Garcia-Sanchez, A., Ania, C.O., Parra, J.B., Dubbeldam, D., Vlugt, T.J.H., Krishna, R., Calero, S.: Transferable force field for carbon dioxide adsorption in zeolites. J. Phys. Chem. C **113**, 8814–8820 (2009)
23. Garcia-Sanchez, A., Dubbeldam, D., Calero, S.: Modeling adsorption and self-diffusion of methane in LTA zeolites: the influence of framework flexibility. J. Phys. Chem. C **114**, 15068–15074 (2010)
24. Loewenstein, W.: The distribution of aluminum in the tetrahedra of silicates and aluminates. Am. Mineral. **39**, 92–96 (1954)
25. Madison, L., Heitzer, H., Russell, C., Kohen, D.: Atomistic simulations of CO_2 and N_2 within cage-type silica zeolites. Langmuir **27**, 1954–1963 (2011)
26. Selassie, D., Davis, D., Dahlin, J., Feise, E., Haman, G., Sholl, D.S., Kohen, D.: Atomistic simulations of CO_2 and N_2 diffusion in silica zeolites: the impact of pore size and shape. J. Phy. Chem. C **112**, 16521–16531 (2008)
27. Goj, A., Sholl, D.S., Akten, E.D., Kohen, D.: Atomistic simulations of CO_2 and N_2 adsorption in silica zeolites. The impact of pore size and shape. J. Phys. Chem. B **106**, 8367–8375 (2002)

MIX
Papier aus verantwortungsvollen Quellen
Paper from responsible sources
FSC® C105338

If you have any concerns about our products,
you can contact us on
ProductSafety@springernature.com

In case Publisher is established outside the EU,
the EU authorized representative is:
Springer Nature Customer Service Center GmbH
Europaplatz 3, 69115 Heidelberg, Germany

Printed by Libri Plureos GmbH
in Hamburg, Germany